Honey Bee Vet

The adventures of a veterinarian seeking to doctor one of the world's most important animals.

Tracy Farone

Northern Bee Books

Honey Bee Vet

Copyright © Tracy Farone

All rights reserved. No part of this publication may be reproduced, stored in a retrieval system, transmitted in any form or by any means electronic, mechanical, including photocopying, recording or otherwise without prior consent of the copyright holders.

Published 2024 by
Northern Bee Books,
Scout Bottom Farm,
Mytholmroyd,
West Yorkshire
HX7 5JS (UK)
Tel: 01422 882751
Fax: 01422 886157
www.northernbeebooks.co.uk

ISBN 978-1-914934-81-0

Design and artwork DM Design and Print

Author Bio

Dr. Tracy Farone, DVM, is a Professor of Biology at Grove City College. Dr. Farone received her DVM from the Ohio State University in 1999. She worked in various areas of private practice for 12 years. She served as medical director for a veterinary technician program and an adjunct professor for Geneva College and Penn State University. In 2010, she joined the Grove City's faculty. She currently teaches a wide variety of courses, and she is active in pre-health advising and student research.

Before bees buzzed into her life, Dr. Farone focused her research on public health, infectious diseases, and ticks. This research led to the collection of nearly 3000 ticks across Pennsylvania. Her publications include three papers on ticks and tick-borne disease. One of these publications, in which Dr. Farone collaborated with the CDC and the PADH, documented the first evidence of Powassan virus found in *Ixodes scapularis* in Pennsylvania.

Since late 2016, Dr. Farone has been researching and re-educating herself in beekeeping and bee medicine. In 2018, she was granted a sabbatical to allow additional time to pursue apicultural studies and develop a small teaching and research apiary at the College. In 2019, Dr. Farone worked intensely in the field with dozens of backyard, sideline, and commercial beekeepers in the US. She traveled to Montana/Crow Reservation to work with large scale, migratory, commercial beekeeping operations. Dr. Farone also traveled to France, where she worked with multiple experts in bee medicine and research at ONIRIS College in Nantes and the OIE in Paris. Additionally, she visited The University of Edinburgh and the Roslin Institute in Scotland, meeting with additional bee experts. Over the past five years, Dr. Farone has had the opportunity to observe the inner workings of thousands of hives. These experiences provided Dr. Farone with a unique perspective in the development of relationships between veterinarians and beekeepers.

Dr. Farone has discovered that bee vet interest is buzzing, and she has been working with multiple veterinary continuing education providers to develop continuing education curriculum for veterinarians. She has given and continues to give bee lectures at multiple universities, veterinary associations, and various bee clubs. She has published several articles on bee medicine, including a "Bee Vet" series for *Bee Culture*, written biosecurity industry guidelines for veterinarians entering bee yards, a chapter in *Veterinary Clinics* Honey Bee Medicine. and developed an educational website, https://www.gccbeeproject.com/ .

Dr. Farone and her research students have built and manage a research and teaching, certified pollinator friendly, apiary garden on the GCC campus. Dr. Farone's work has been featured in the *JAVMA*, and she is on the Board of the Honey Bee Veterinary Consortium and the American Veterinary Medical Association's Agricultural Animal Liaison Committee representing honey bees. In her free time, Dr. Farone enjoys spending time with her family, running, horse-back riding, SCUBA diving, and of course, just "beeing" with her backyard hives.

Forward/Endorsements

"Tracy m'avait prévenue de son projet, mais je n'avais pas compris l'angle qu'elle choisirait pour écrire sur cet insecte aussi attrayant que mystérieux.

Le sous-titre nous en dit plus, on sent déjà de l'admiration de la part d'un docteur vétérinaire apicole, mais aussi la capacité à s'interroger sur la difficulté pour un vétérinaire praticien classique (d'animaux vertébrés) à soigner une colonie d'insectes. Nos abeilles vivant en colonie très organisées ont un langage différent de la grande diversité de vertébrés que nous tous, vétérinaires avons appris à comprendre durant nos études et notre pratique.

Lorsqu'elle nous a rendu visite en 2019, à l'École Nationale Vétérinaire de Nantes (Oniris), nous avons échangé sur la mise en place de notre enseignement post-universitaire de la pathologie apicole et sur l'implication des vétérinaires de terrain, en France. Ayant moi-même fait ce long chemin pour impliquer les vétérinaires dans la pratique apicole, je mesurai l'ampleur de la tâche qu'elle souhaitait engager.

Ce livre se lit comme un livre de contes, alliant pédagogie, connaissances et humour. Les références récurrentes à l'anatomie, la physiologie, la pathologie (etc.) comparées par rapport aux animaux vertébrés est remarquable. Son analyse relève d'une réflexion profonde que doit effectuer tout vétérinaire confronté à une espèce animale autre que son modèle habituel.

En effet, les vétérinaires ont appris une démarche diagnostique qui leur ait devenue intuitive et, ceci quelle que soit l'espèce à laquelle ils sont confrontés. Tracy nous fait comprendre que, pour peu que les vétérinaires aient les bases anatomique, biologique, etc. et qu'ils aient connaissance de la pathosphère d'*Apis mellifera*, ils sont à même de comprendre la gestion intégrée des problèmes pathologiques apicoles. C'est le premier pas du vétérinaire apicole...

Merci Tracy pour cet éclairage !"

"Tracy had warned me of her project, but I hadn't figured out what angle she would choose to write about this insect that was as attractive as it was mysterious.

The subtitle tells us more, we already feel admiration on the part of a veterinary beekeeper, but also the ability to question the difficulty for a classical veterinary practitioner (of vertebrate animals) to treat an insect colony. Our highly organized colony bees have a different language from the great diversity of vertebrates that all of us veterinarians have come to understand during our studies and practice.

When she visited us in 2019, at the National Veterinary School of Nantes (Oniris), we discussed the implementation of our post-graduate teaching of beekeeping pathology and the involvement of field veterinarians in France. Having made the long journey to involve veterinarians in beekeeping, I realized the magnitude of the task she wanted to undertake.

This book reads like a storybook, combining pedagogy, knowledge and humour. The recurrent references to anatomy, physiology, pathology (etc.) compared to vertebrate animals is remarkable. Its analysis is the result of an in-depth reflection that must be carried out by any veterinarian confronted with an animal species other than his usual model.

Indeed, veterinarians have learned a diagnostic approach that has become intuitive to them, regardless of the species they are confronted with. Tracy makes us understand that, as long as veterinarians have the basics of anatomy, biology, etc. and that they are familiar with the pathogenosis of *Apis mellifera*, they are able to understand the integrated management of beekeeping pathological problems. This is the first step of the beekeeping veterinarian...

Thank you, Tracy for this insight!"

Dr. Monique L'Hostis, well-published author and researcher, bee veterinarian, developer of Honey Bee Program at the veterinary school of Nantes, France, and my mentor

"This book by a longtime practicing and teaching veterinarian offers fresh insights into both beekeepers and veterinarians curious to know more about biology and the culture of healing. The book has a clear format based on the sections (frames) of a beehive and covers all of the basics of bee biology and bee challenges while also adding the fresh perspectives of someone who knows both fields well. Veterinarians will benefit from clear prose describing bee biology, distributions, and the longstanding tie between humans and honey bees. The 'frames' (5 and 6) covering hive inspections and disease management give an organized and fresh take on these topics, showing the value of setting criteria for investigating scenes and making rational decisions. These sections will be unique for many beekeepers and will improve their hive management, as will the closing discussion of biosecurity, an overlooked component of maintaining apiary health. The author is a clear and engaging writer, and the book delivers insights in an active and readable form."

Dr. Jay Evans, Lead Scientist USDA-ARS Bee Research Laboratory, Beltsville MD

"Dr. Tracy Farone discovered Honey Bees unexpectedly like we all have. But she is a veterinarian, and a college professor so her journey has been the same but different. I met Tracy and knew that her knowledge, skills, and abilities would be so connecting, fun, and insightful, for Bee Culture readers as 'Bee Vet'. Now, she has collected her Bee Vet articles into a very understandable, and practical book to help all of us be better Beekeepers. Which means our relationship with Honey bees will be even stronger and more successful. Her book is inviting, readable information. Great job!"

Jerry Hayes, Bee Culture Magazine Editor

Dedication page

To my husband and son. You are the gifts of my life, my cheerleaders, always supportive and patient...an ever-present rock that provides love, stability, sanity, and humor in this crazy world. Thanks for putting up with bees everywhere in our lives! I am so thankful to the Lord and blessed to have you in my life. I love and cherish you both.

What is in Stores

Frame One: At the Bee-ginning

Introduction	1
Festooning: The Importance of Community and Action.	2

Frame Two: Honey Bee Biology

Sentinels	7
When Considering the Data	10
The Hexagon, Under the Microscope	14
AI in Beekeeping?	17

Frame Three: Honey Bee Anatomy and Physiology

Time for Some Muscle	22
All Things Change	26
New Beginnings	33
In with the New and Out with the Old	35
Bee Beats and Breaths	41

Frame Four: Honey Bee Behavior

Human-Bee Bond?	45
How to Stop Swarming	50
The Conundrum of Feral Honey Bees	58

Frame Five: **Honey Bee Management**

The Veterinary Diagnostic Approach:	*63*
*The Exam	*65*
*Finding the Diagnosis	*66*
*Treatment	*70*
If We Only Followed the Directions!	*74*
Fall Planning	*79*
A Surgical Approach	*83*

Frame Six: **Honey Bee Medicine**

Diagnosis Is a Tricky Thing	*90*
Euthanasia	*92*
Immunity, Vaccines, & Honey Bees	*97*

Frame Seven: **Honey Bees Diseases**

Tropilaelosis	*103*
Not as fun as a Volkswagen	*107*
The Diagnosis of the Shrew	*113*
Emerging Diseases in Honey Bees	*115*

Frame Eight: **Honey Bees, Humans and Public Health**

"Good" Help	*121*
Bees in Christendom, the Spiritual Health of Honey Bees	*124*
Biosecurity for Beekeepers	*127*
Another Arachnid	*131*

The Super

Acknowledgements	*138*
Notes	*139*

Frame One: At the Bee-ginning

Unexpected journeys often lead to the best adventures. In this 8-frame bordereau, I would like to share the experiences of my unexpected journey into the world of apiculture. Over the last seven years, I have been blessed to spend countless hours listening, observing, and working in the field with dozens of beekeepers, entomologists, bee researchers, and other veterinarians. Topics pondered will include the latest "buzz" on our common industry challenges and solutions, what vets can do for honey bees and beekeepers, and highlight topics in biology, pathology, diagnostics, and treatments related to honey bees. This first "frame" will give you a bit of background into how things all got started in connecting or "festooning" the worlds of beekeepers and veterinarians.

Getting Ready for Flight

Festooning: *The Importance of Community and Action.*

When I heard about veterinary bee medicine in the US for the first time in late 2016, I laughed. I was seventeen years into my veterinary career and thought I had heard it all. Nope. Bees...(giggle). Even with a background in tick and public health research, I initially thought the idea was **just one more thing to do**. I tried to dismiss the idea. But have you ever had something just keep "bugging you", until you checked it out? I started reading. Reading anything and everything I could about honey bees and honey bee medicine. I uncovered connections and discovered that my College (of employment) had a relationship with a veterinary school in Europe which boasted a post-doctorate program in bee medicine. I also discovered that my host family at the Crow Indian Reservation in MT (a trip I took every year with students) had connections with large 10K hive migratory beekeepers. I was intrigued.

Research students next to bee shed

I realized that bees have bacterial, viral, parasitic, fungal, nutritional, and toxic diseases, like other animals. Bee diseases can be diagnosed through exam findings and diagnostic testing, and treated with environmental, nutritional, and pharmacological interventions, which are medical approaches taught universally in veterinary schools.

I was also concerned to learn about the many challenges honey bees, other pollinators, and beekeepers face. I wanted to **learn how to help**, as it became apparent that I was in a unique position **to do** something. But I needed to understand beekeeping first and eventually become a beekeeper. I needed prep. I applied for and was awarded a sabbatical to study this new quandary of bee medicine coming to the US. So since late 2016, I have been on a journey spending thousands of hours studying the evolution of this new relationship and creating an apicultural research program at my College. This journey, I would like to share, and continue, with you.

In much of Europe, an origin of our beloved *Apis melifera* , honey bees have had a doctor, just like any other agricultural or companion animal, for decades. Bee veterinary medicine is a typical course of instruction at veterinary schools, taught alongside all the other species veterinarians may serve: cattle, hogs, chickens, horses, dogs, cats, birds, sheep, goats, rabbits, etc. An apiary is an integral part of the students' educational rotations on European veterinary campuses. It is the norm for veterinarians to be educated and contribute to the care of arguably our most important agricultural animal, the honey bee.

During my sabbatical, I traveled to France to visit ONIRIS veterinary college, (the one that had the post-doc bee program). In the semester prior to leaving for Europe, I thought it would be a good idea to learn some French. So, on my daily two-hour commute to and from work, I listened to a French conversational podcast in an attempt to learn the language. Now, I do speak some French... badly. Once in France, I worried I would insult my French colleagues by butchering their beautiful language. **But I tried.** To my surprise, my attempt to learn their language was the key to opening up real relationships and trust. They welcomed me like family into their homes. We shared many meals together, and they shared what was personal and important to them. My French colleagues took me **all over** France showing me large portions of their bee industry, from commercial beekeeping to honey and wax processing, to their bee veterinary medical curriculum. Explaining much of it in French! I was overwhelmed. I still have colleagues who have become friends, which I correspond with regularly...in French.

After France, I spent some time in Scotland (sorry Outlander fans, I did not find Jamie Frazier). Scots take pride in doing things the way they like to do things, including beekeeping, but there is no real language barrier here, right? Well...while walking the Royal Mile in Edinburgh, I wandered into a kilt shop and found a Mackenzie gentleman who took the time to tell me about every tartan's family meaning, colors, and history for at least 45 minutes. I listened smiling attentively, but he once kindly asked if I "dinna" understand what he was saying. I dinna always, but I got the gist and greatly appreciated his effort.

Edinburgh Castle

Just outside of Edinburgh, I visited the University of Edinburgh, which boasts the Roslin (research) Institute, famous for the first cloned animal, "Dolly", the sheep, and the Royal Dick School of Veterinary Studies. Yes, Royal Dick. Language, remember... some things mean different things to different people, from different perspectives. I was taken on a wonderful tour of the facilities by local gentlemen who proudly told me all about the School's history, and the contributions and shenanigans of their founders, William Dick and more likely, William's sister, Mary Dick.

Of course, with the veterinary school, there was also a bee yard and beekeeper/bee researcher who donned a very impressive beard. The apiary was hidden in plain sight, through a sheep field filled with perhaps Dolly's colleagues? Clones? While studying in the apiary, I noticed locked ratchet straps around the hives. I knew there were no bears in Scotland but thought maybe there may be other pests. So, I asked the beekeeper about the locks. He said with a wink, "Awck, we dinna have bears in Scotland, but there **are** other Scots." Despite all the science I learned about bees in Europe, my take home lesson was that to learn the **language** of another group or person is the key in developing successful relationships.

Back in the States, I had a beekeeper, now a dear friend of mine, tell me the importance of being a "doer". Perhaps because he initially thought I was a stuffy academic, but as a clinician with 12 years in the field, I understood what he was saying, took no offense, and then helped him physically build a commercial bee yard. Bees are certainly doers – they get things done. Language is important not just to speak, but to listen and ignite mutual understanding and cooperation.

Festooning is a funny little bee word. It sounds like a mix between festival and cartoon. In bee biology, festooning is a very important piece of language to understand. Festooning, regarding honey bees, is defined as a behavior in which bees cling to each other, often in single chains, reaching out their limbs to each other to make connections, with the intent to **build the framework of something new.** Sometimes the behavior occurs **to repair** old comb or measure **distances between spaces**. Sometimes scientists are not sure why they do it. If you are a beekeeper, you have probably noted this behavior with a smile. It is fun to watch.

Frame Two: Honey Bee Biology

In order to understand the nature of any beast, one must become familiar with how it is made and how it operates in our world. This is the study of Biology. The study of living things within our unique environment of Earth. Please do not be intimidated by the "scientific" sound of this but understand biology as a love that seeks to know creation. This love must be the first love fascination of any beekeeper hoping to fully understand their honey bees. This is why chapters featuring honey bee biology stories are presented at the beginning of this book, as a foundation of understanding what's to come.

Honey bee on golden rod

Sentinels

It can be amazing how things seemingly unrelated are ultimately connected. Perhaps that is one reason why honey bees continue to fascinate me, as they so often demonstrate the story of One Health relationships.

In January 2019, I had the privilege of visiting Notre Dame Cathedral in Paris... just a few months before the infamous fire of April 15th, 2019. In Europe, there are many lovely churches and cathedrals, but Notre Dame is a true standout. Beautiful, awesome, peaceful, and intimidating all at the same time. When I visited, the place was still decorated for Christmas and the organist was phenomenal, especially if you can appreciate a creepy "Phantom of the Opera" type of organ tune. Ironically, at the time, I was also in France for the purpose of studying apiculture.

Last Christmas of Notre Dame

The cathedral's three honey bee hives are now famous in the bee world for "surviving" the fire. The fire, in addition to partially destroying the magnificent building, was also considered to be an "acute pollution event". Among other things, the ancient structure contained considerable amounts of lead. Following the fire, studies were conducted utilizing honey samples from various parts of Paris to measure lead levels. Honey samples directly downwind from the fire had two and a half times the lead of pre-fire levels, three times the levels compared to other parts of Paris, and six times the lead levels comparable to the wider geographical region. The good news is that even though post-fire, downwind, lead levels showed elevation, the levels in the honey were still considered safe for human consumption. Most importantly, the information the honey bees and their honey provided was a useful bio monitor of the environmental contamination of the area, helpful in assessing health risks for humans (1).

It is well documented that bees can be utilized as sentinels for various environmental pollutants including various heavy metals (such as lead, cadmium, chromium, nickel), pesticides, and other chemical substances (2, 3, 4). In medicine, the word **"sentinel"** is a reference to something that indicates the presence of a disease or health threat in an area. We often study various animal sentinels in veterinary medicine and can utilize this information to recognize and analyze human disease risk. For example, testing ticks for various infectious diseases and/or clinical testing of canines for Lyme Disease antibody can indicate disease exposure risk to humans and animals in a geographical region.

In the US, there is little found in the literature regarding heavy metal exposure affecting honey bees clinically and acutely. In Europe and elsewhere, there are some studies pointing to heavy metals as having detrimental clinical effects on honey bees, particularly in historically contaminated areas (5,6). It is well established that honey bee bodies can and do pick up metal levels from the environment through air, water, and plants (2-6). On most days, low level heavy metal exposure is one of many environmental pollutants our bees must endure which **may** factor into difficult to diagnose subclinical manifestations. I would argue however, that if a "acute pollution event" occurred in your area, heavy metal toxicity could be a rare but possible differential consideration.

The take away from all of this is that for many things, from chemicals to infectious agents, it is not so much the "what" but the "how much" that determines effect...the dosage. While we cannot control all the ever-present contaminants in our bees' foraging range, we can control what is in our bee yards. The place our bees spend the bulk of their life.

The common definition of "sentinel" is a guard who stands and keeps watch. So, while bees can act as medical sentinels for us, we can act as sentinels, watchful keepers, for our bees. Every chemical, every metal, every substance that you permit in your bee yard, are in the closest proximity to your bees most of the time. Particularly the substances we put **on** our bees, like medications, including antibiotics and miticides. Appropriate and judicious use of such products is a key component to being the best stewards of our honey bees.

When Considering the Data

If I asked you, "What is normal human body temperature?", what would be your answer? Most people (and maybe even some health professionals) would say 98.6 degrees Fahrenheit (or 37 degrees Celsius). You may have solidly believed this your whole life. However, this data point is based on one, single, now highly disputed study performed by a German physician, Dr. Wunderlich, in the mid-1800s, taking axillary (armpit) temps with a questionable thermometer (1,2,3). In psychology, they refer to this as the "illusory truth effect" or if something "is repeated enough times, the information may be perceived to be true even if sources are not credible." (4).

As a veterinarian for over 20 years now, I always wondered about this single, 98.6, human number because, as vets, we were not taught single values for normal animal body temperatures but **ranges of normal**. For example, normal horse temps are from about 99-101.5F with anything over 102 raising our eyebrows. Cattle average about 101.5 with anything over 103 considered to be a fever. Cats and dogs fall into a 99.5-102.5 range, but with freaked out cats coming into a clinic, I'm not surprised by a 103 or even higher given the situation. As thermoregulators, even our honey bees are capable of a wide range of "body" temperatures within their colonies. Brood and winter core clusters temps range from 90-95 F, with outer parts of the winter cluster being much cooler, with ranges from 81 F to the mid-40s (5).

Body temperature *ranges* make sense because body temperature is a **variable** that can fluctuate and change within a *normal homeostatic range* depending on host and environmental characteristics. Body temperature in individual animals and humans can be affected by many things, including age, height, weight, activity, gender, stage in reproductive cycles, pregnancy, and the time of day.

The accuracy of measuring this variable also depends on the use of *proper instrumentation and methodology in our data collection*. Regarding body temperature, internal testing methods, rectal and oral thermometers are the most accurate for measuring body temperature (we tend to use rectal thermometers in young children and animals and oral in older children and adults). External, non- contact thermometers, which are now popular due to convenience and perceived safety, are less accurate. We also need to be sure our method is correct. Remember mom saying, "be sure to put it under your tongue"? She was right. Improper placement, inadequate testing time, being outside, exercising, eating, or drinking before using an oral thermometer will affect its accuracy (6).

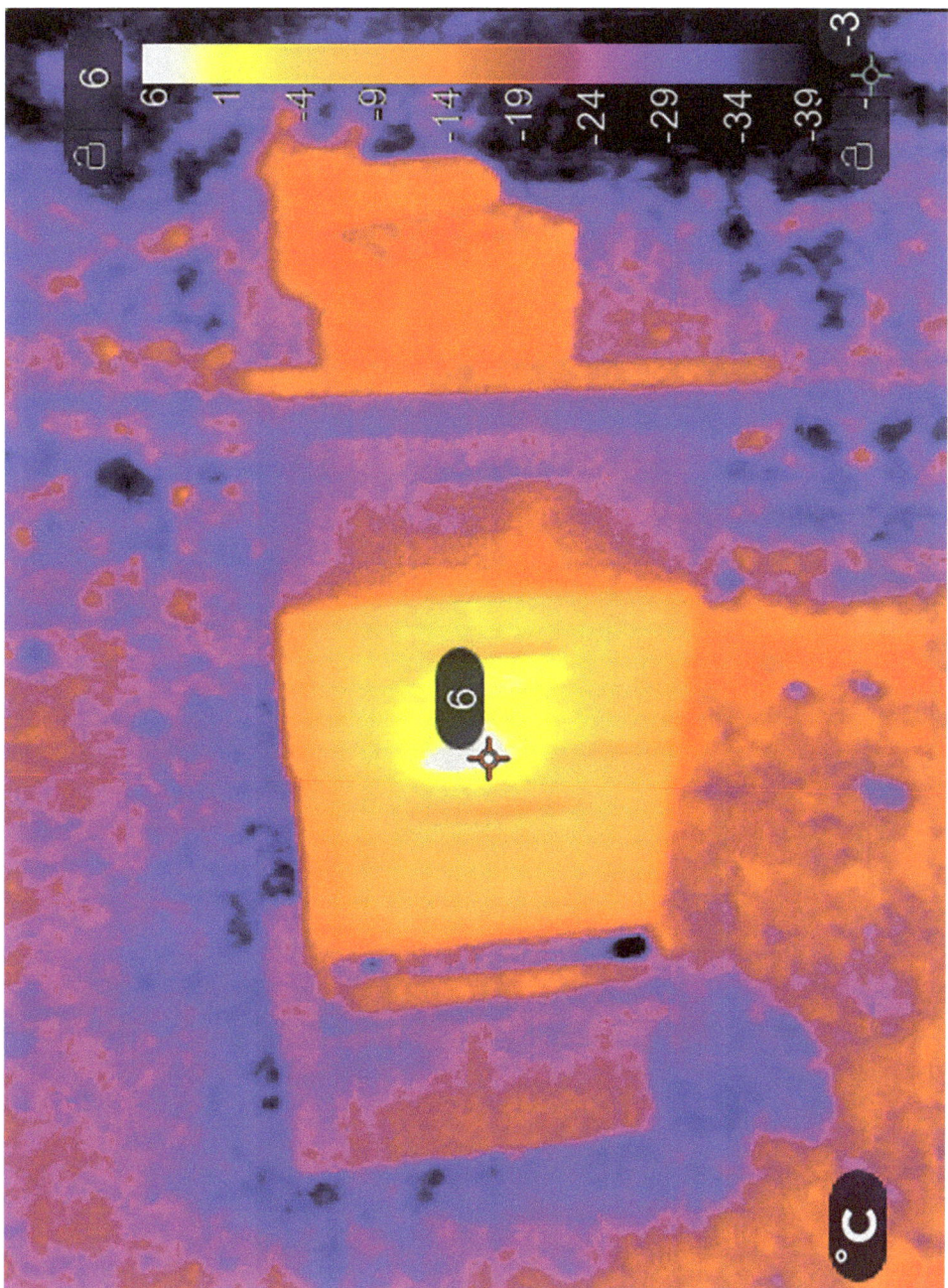

Bees' warm cluster

The current scientific consensus involving many more recent studies on human body temperature, (which managed some main stream press coverage in pre-pandemic 2019), now considers humans to have a normal range of body temperature somewhere between 96-99.5 F, with an average of about 97.7 F. Most doctors agree that 100.4 or above is definitely a fever (personally, I'm wiped out by anything over 99 F). Does this surprise you?

I present to you this "temperature check" to illustrate that in any science, it is very important to not just know the "facts" or data presented, but to be able to fully ascertain what, how, when, where, and why in the generation of the information. Once information is elevated to "facts", this information becomes what we base our beliefs and actions on. It is also important to accept that for many things there may not be a single, yes or no answer but a range of "correct" answers affected by many variables. Much of what I presented above is analogous to how we should evaluate data used in beekeeping decision making.

Recently, I was asked to speak at a local beekeeping club on small hive beetles. I enjoy the opportunity to speak with beekeepers because I try to not be the only talking head in the room and give the group the opportunity to share their thoughts as well. It gives me great food for thought when listening to other people's perspectives and you never know what others are thinking unless you ask. At this particular meeting, I was asked, "Are (all) treatments for Varroa mites effective in treating hive beetles?... Yes or no?!!"

The short answer to this question is, "No."

The label is the law. Miticides for use on honey bees are not formulated, intended, or labeled to kill an insect like small hive beetles. But I believe this beek may have been wanting me to say, "yes", because we know that many chemicals (insecticides, pesticides, fungicides, herbicides, miticides) can potentially have negative, even if sub-clinical, effects on insects, like our bees, so why not small hive beetles? Maybe we could *assume, believe,* that there is some effect on hive beetles, too...? I suppose if you'd put a hive beetle or just about any other living creature into a vat of amitraz, oxalic acid, or thymol for a time, it would not be promotional to their good health, but this of course, would not be anything we would do in practicality or be backed by any scientific study that I am aware.

However, after the meeting, I got to thinking. While the technical answer to the posed question is, "No", the IPM (Integrated Pest Management) answer to the question is – Yes! If beekeepers are properly using an effective Varroa treatment program and controlling mites in their hives, they are doing one of the most important things beekeepers can do to keep their hives strong with effective immunity. And keeping strong, healthy hives is one of the best preventions in fighting small hive beetle infestations. So, yes.... the range of "correct" answers depends on your perspective.

Everyone likes to be correct...to know what they are talking about. But science and beekeeping has continued to teach me to be diligent in considering questions and answers, the credibility of data resources, and to understand that there are *many variables* going on at the same time, particularly in the field with wild animals such as honey bees. Additionally, if one investigates just about *any* topic beyond a typical "Google" search, you'll shortly find that even for "experts" in a particular field, we're honestly just a few questions away from, "I don't know....much is unknown... the data is changing... or more research is needed", before we can jump to any one-size-fits-all conclusion.

The Hexagon, Under the Microscope

Each Fall for about 10 years now, I have had the joy of teaching a course called "Histology". Histology is the study of cells, often with the aid of a variety of microscopes. It is not a sexy course title, but the course content is relaxed and involves looking at a lot of colorful pictures. My students spend hours peering through the lens of a microscope trying to visually decipher the blobs before them. By the end of the course, after enough staring, most students achieve the skill of identifying "the blobs" as some specific cell or tissue. Since the late 16th century, the study of histology has given humans the opportunity to give things a "closer look", and the technology is ever increasing our ability to "see" beyond the cellular level.

Histological identification certainly has many practical applications in biology, health care, disease diagnosis, and in beekeeping. Closely examining our honey bees' *body cells* has led to the understanding of their functional anatomy, physiology, pathological tendencies, and disease diagnosis. What I find incredibly fascinating is the link between the shape of our honey bees' *comb cells* and what is also found microscopically in many of our body's tissues. The hexagon.

I am sure you have read articles about this shape, which is so often associated with the beautiful honey comb pattern of our bees, and how hexagons are the most effective and efficient geometric shape. These 6-sided polygons can tessellate any plane to completely fill the area with no gaps. From art to architecture, in snowflakes, soap bubbles and soccer balls, natural design or man-made, hexagons are everywhere, appreciated as a symbol of strength, stability, and security. But have you heard about the hexagon from a histologist's perspective? Well, allow me to give you a different take. Here is today's lesson:

Common types of microscopes and honey bee applications

1. Dissection scopes: Dissection scopes are commonly used to view smaller, 3-D things (like bees) with up to a 100x magnification. These scopes are helpful in studying plant and insect morphology, and for taking "zoomed- in" photographs.

2. Optical light microscopes: These scopes are probably what you imagine when you hear the word "microscope". Using visible light and a series of lenses, these scopes typically can magnify objects up to 2000x. Examined specimens are typically prepared onto slides for viewing. These microscopes are invaluable in diagnosing a variety of abnormal cellular formations and infections in man or bee, including parasitic, bacterial, and fungal diseases.

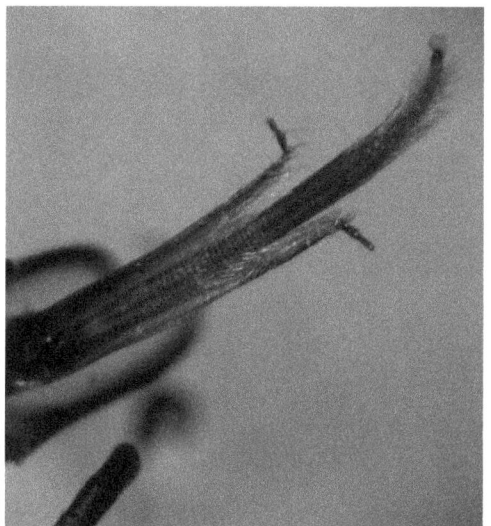 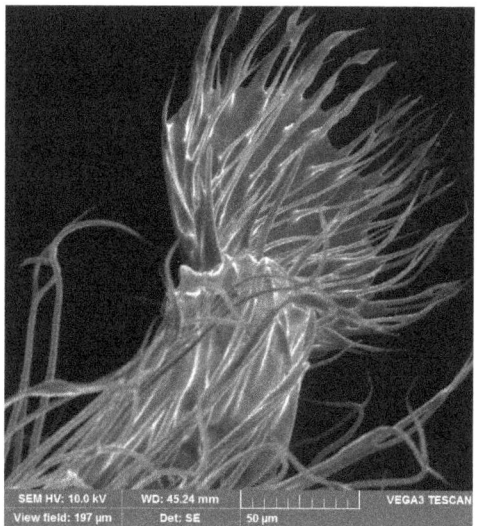

Dissection scope view of a honey bee   SEM showing the tip of a honey bee

3. Electron microscopes (EM): Electron microscopes actually use a beam of electrons (not light) to create images magnified millions of times. They are typically found in research labs and universities. There are two major types.

   a. Scanning Electron Microscope or SEM: Scanning electron microscopes create a sharp almost 3-D like image of the surface of a specimen. Magnification ability of a SEM can reach one-half a million times. We have a SEM in our lab that my students use to capture cool images of honey bees and their parasites.

   b. Transmission Electron Microscope or TEM: Transmission electron scopes capture a flattened slice of tissue magnified up to fifty million times. TEMs can be useful in diagnosing and studying pathogens.

4. Fluorescent microscopes, FS: Fluorescent scopes utilize fluorescent dyes added to a specimen to tag certain tissue components or pathogens. When exposed to UV light, these fluorescent dyes absorb lower wavelengths of light and then emit a higher wavelength or seem to "glow." If the examined sample glows appropriately, we know that we have found what we are looking for...amazingly, there are now small, portable FS that work with a smart phone to detect Nosema in honey bee samples in the field. Yep, there's an app for that!

Histological Hexagon Examples

Many of our body's cells are designed in a hexagonal mosaic. Chemists could even argue that at a molecular level, all carbon-based life forms contain a base of hexagonal carbon rings, but I am not a chemist, so let us just look at what we can see (through a microscope).

The urinary bladder, for example, has a unique and specialized ability to expand and contract over short periods of time. This can come in handy when you have had that extra cup of coffee! This peculiar capability also comes with a unique design. As with most things in nature, form and function are always complementary. The bladder has a specialized type of tissue that lines its interior called **transitional epithelium**. It is called transitional because these cells of the bladder are shapeshifters. They range from almost flat to taking on the appearance of an overstuffed marshmallow. When the bladder is full, the cells spread out and take on the shape of, you guessed it, the hexagon. Even a vessel that is given the lowly task of holding urine has a beautiful design from the inside out.

The liver is the second largest organ in most mammals and performs thousands of functions necessary for the preservation of the body. (Even honey bees have an organ that mimics the important work of the liver, in their fat bodies). The liver is arranged in rows of cells, stacked in columns forming a sieve in which blood can slowly flow through the organ, so that each drop can be processed by the liver cells, or hepatocytes. What shape to best accomplish such a feat? Yep, again - the hexagon. Each hepatocyte itself is a hexagon, which then combines in groups to form larger units called lobules, also hexagonal, collectively, in shape. Again, the hexagon is the best way to get most things done in the best use of space.

Other 6-sided cell examples in histology include the inside surface in much of the small intestine, which takes on the appearance of hexagons with a fuzzy border of projectiles, called microvilli. These shaggy microvilli help to increase surface area to facilitate the absorption of our latest meal. Another example are insects' eyes, including our honey bees' eyes, these compound eyes are made of thousands of individual eyes arranged in a hexagonal mosaic pattern. Even the very fibers of our own eyes' lenses are composed of long layers of hexagonal cells. Hmmm, ironically, I suppose we are looking at hexagons through a bunch of hexagons.

If you look around, you may be able to find a dissection or optical scope on the cheap to play around with and get a closer look at your bees. Since I have become a beekeeper, I have yet to find a subject that does not somehow relate back to bees or the study of them.

AI in Beekeeping?

"Greetings beekeeping enthusiasts! Today, we'll dive into the world of artificial intelligence and explore its application in the beekeeping industry. Specifically, we'll discuss ChatGPT, an innovative technology developed by OpenAI, and how it can help promote honey bee health and management.

First things first, let's define what ChatGPT and OpenAI are. OpenAI is a research organization dedicated to developing advanced artificial intelligence in a safe and beneficial way. ChatGPT, on the other hand, is a language processing model developed by OpenAI. It uses deep learning algorithms to analyze vast amounts of natural language datoa and generate responses that closely resemble those of human conversation.

Now, you might be wondering how ChatGPT can be useful in the beekeeping industry. Well, let's take a look at how AI has already been used in this field. In recent years, AI has been used to develop sophisticated monitoring systems that can track various aspects of bee colony health and behavior, such as population size, hive temperature, and humidity. These systems have been instrumental in identifying potential problems in the hive before they become serious, allowing beekeepers to take preventive measures.

But what if beekeepers could access all this information and more simply by asking a question? This is where ChatGPT comes in. By using natural language processing, ChatGPT can be programmed to understand and respond to beekeepers' queries in a way that mimics human conversation. For instance, a beekeeper could ask ChatGPT about the health of a specific colony or inquire about the best way to treat a particular bee disease. ChatGPT could then provide a response based on the latest research and data, helping beekeepers make informed decisions about the health and management of their hives.

However, as with any technology, there are pros and cons to consider. One of the advantages of ChatGPT is that it can save beekeepers time and money by providing quick answers to their questions without the need for extensive research. Additionally, it can improve the accuracy and consistency of the information beekeepers receive, as ChatGPT's responses are based on objective data rather than personal opinions.

On the other hand, there are concerns that relying too heavily on AI could undermine the importance of hands-on experience and intuition in beekeeping. Additionally, there are issues surrounding data privacy and security that must be addressed to ensure that beekeepers' information is protected.

Despite these concerns, it's clear that AI has the potential to revolutionize the way beekeepers manage their hives. By combining AI with traditional beekeeping practices, we can achieve a more sustainable and efficient approach to honey bee health.

In conclusion, ChatGPT is a cutting-edge technology that has the potential to transform the beekeeping industry. Its natural language processing capabilities can help beekeepers access vital information about the health and management of their hives in a quick and convenient way. However, as with any technology, it's important to weigh the pros and cons and ensure that it's used responsibly and ethically. With the right approach, AI can help us achieve a more sustainable and successful future for honey bees and the beekeeping industry as a whole."

If you made it this far, congratulations! You were reading the ramblings of a computer. What did you think? Are you surprised? The above text was generated with me working with ChatGPT utilizing the following prompt.

"Write up to a 2000-word article starting with: a brief introduction and definition of what ChatGPT and Open AI are, what ChatGPT can do, and the pros and cons of this technology. Then give an overview of how AI has been used in the beekeeping industry and how it may be used in the future to promote honey bee health and management. Write the entire article in the style of Dr. Tracy Farone, author of the "Bee Vet" column in Bee Culture Magazine. Dr. Farone's writing style is conversational but scientific. Do not introduce Dr. Farone in the article, the audience will know who she is. Do not end the article without completing a sentence but finish with a conclusion."

This was the eighth prompt and attempt of Chat to get to a writing of this quality. Previous times the instructions were not followed, the article stopped in midsentence, information given was superficial, and the tone was cheesy car salesman at times. The word count is still only 517.

In the above computer-generated writing, as a critical human thinker, I believe one can see how much Chat thinks of itself- it's quite the happy narcissist! It claims to be objective, but I think the biggest worry is "garbage in, equals garbage out". The potential for bias, dangerous bias, is a large concern. However, objective truth is anything but simply objective anymore. True science is about the ability to ask questions and critically reason through them. It is the line between AI and self-awareness...

There is a hazy line defining the future of AI use in

Even though ChatGPT has access to and the ability to compile millions of data sources, Chat left out a few things in its article. It did not discuss automated beehives in detail (from Beewise- Beehome) or various in-hive monitoring devices or apps considered to be forms of AI. I find anything that could make improving hive health to be well worth a try but, of course, it must practically work and make economic sense as well.

Many are praising the technological capabilities and applications of this new birth of AI, considering it is just something we will have to adjust to like the Internet and social media (Has the pros outweighed the cons here?). Others, including key developers like Elon Musk, are calling for a pause in this rapid pace of the development of

technology before we can fully understand what it will do in our world. A few abilities of Chat GPT that are being promoted but also cause concern include:

1. The ability to create malware endlessly. AI does not need to sleep, eat, or take out the garbage.

2. Pretending to be someone (human). For example: dating apps (creeeepy!), better phishing/scam messages, writing resumes.

3. Assisting in cheating in school, college, graduate school, medical schools...

4. Giving advice to the user on relationships, psychological support, even medical advice.

5. Being a friend or companion to the user.

6. Taking over multiple human jobs.

A few weeks ago, I sat in a room full of college faculty listening to our computer science department give a presentation about what ChatGPT/Open AI is and what it can do. I was hoping to get some answers about how we handle this new technology. To be honest, the mood in the room was grim. From a teaching perspective, this technology brings up huge ethical and cheating concerns, as Chat can write original papers for students, do their homework for them, and some claim even write Ph.D.s. Technology to detect such shenanigans is available but unable to keep up with the latest Chat versions developing by the day. How are young minds ever going to develop critical thinking and acquire knowledge if a computer can simply do it for them?

At the meeting, the suggestions we received suggested going back to paper and pencil writing in-class assignments, exams, and oral examinations for true assessments of student knowledge. Great... advancement leads to going back in time. Clearly these methods are not practical in large classes, primary writing courses, or on-line classes. However, after playing around with Chat for this article, I was a bit relieved that, at least for now, those of us humans with functioning brains may still have the upper hand. I will leave you with one last conversation I had with Chat...

Me: How do you get out of a cardboard box?

ChatGPT: As an AI language model, I do not have a physical body, so I cannot get out of a cardboard box. However, if you find yourself stuck in a cardboard box, the best way to get out would be to push or pull the flaps open from the inside or cut the box open with a sharp object. It's important to be careful when cutting the box to avoid injury.

Me: So, a person would use their brain to figure it out.

ChatGPT: Yes, that's correct. A person would use their problem-solving skills to figure out how to get out of the cardboard box.

Frame Three - Honey Bee Anatomy and Physiology

Many of my students dream of becoming medical professionals because they want to help people...learn how to fix them. Beekeepers are the same way. They want to know how to fix their bees if their bees get sick. And they want to be able to do it now! However, the foundational understanding required to recognize abnormal is to first study, in-depth, what normal is. In order to recognize disease, one needs to grasp what health looks like. This is why Anatomy and Physiology are foundational courses in medical schools before any attempt to diagnose and treat (fix) anything comes into play. It is the same in beekeeping. This frame will provide an overview of normal in the honey bee.

Normal honey bee comb

Time for Some Muscle

As I am writing this, it's Labor Day. It's a time of transition in many things, back to school, beekeeping chores, summer to fall. One of the jobs I have been privileged to do for the past 20 years or so, is to teach Anatomy to students who will most likely pursue a career in a health field. It is and has been a great privilege to lead these students through the foundational language and the creative form of their new calling. While human anatomy is largely our focus, I also offer students a taste of comparative anatomy between mammalian species to illustrate the similarities in anatomical design. Anatomy is also paramount in understanding normalcy and pathology in honey bees.

Volumes have been written (and drawn) on the subject of honey bee anatomy, and I will not even try to tread upon the greatness of Snodgrass. Yet anatomy remains the foundational science of medicine, including honey bee medicine. So, I've decided to write a series of mini lessons about these foundational basics for you from a clinical, system by system, in-the-field perspective.

A good beekeeper must know what normal is before one can identify abnormalities in your hives. How do you get good at identifying abnormal? Answer: Look at a whole lot of normal. In this chapters' section, you will receive a free ride to go along with my class. I will break down important anatomical things to look for as well and their medical significance. We will start with the musculoskeletal system, dive into the control systems of the body, the nervous system and endocrine systems, take a peek at the reproductive systems of the bee, move through the gut and excretory systems, and finish with the open vascular and respiratory systems of the honey bee. By your next hive inspections, you'll be ready to apply what you have learned. In a comparative sense, it is the same syllabus for my pre-med students. Let's go!

Skeletal System of the Honey Bee

Form & Function: In honey bees, their outer covering is a combination like that of skin and bone in mammals and humans. This exoskeleton provides an armor-like barrier between the environment and the rest of the body. The exoskeleton provides physical protection, much like skin and bone, but all from the position most exterior to the rest of the body. Protection from dehydration and the entry of pathogens into the body, as well as plentiful hair are more skin-like in nature. A waxy coating covers the exoskeleton. These protections play a large role in innate immunity, the first line of defense for any organism against disease agents.

Much like bone, the exoskeleton provides attachment points and support to the skeletal muscles of the thorax. The color of the exoskeleton can vary in pigment and pattern with yellow, orange, black, and grey hues commonly found in a variety of *A. mellifera* sub-species. Various glands' ducts exit and release their product through or around the exoskeleton, ex. wax glands.

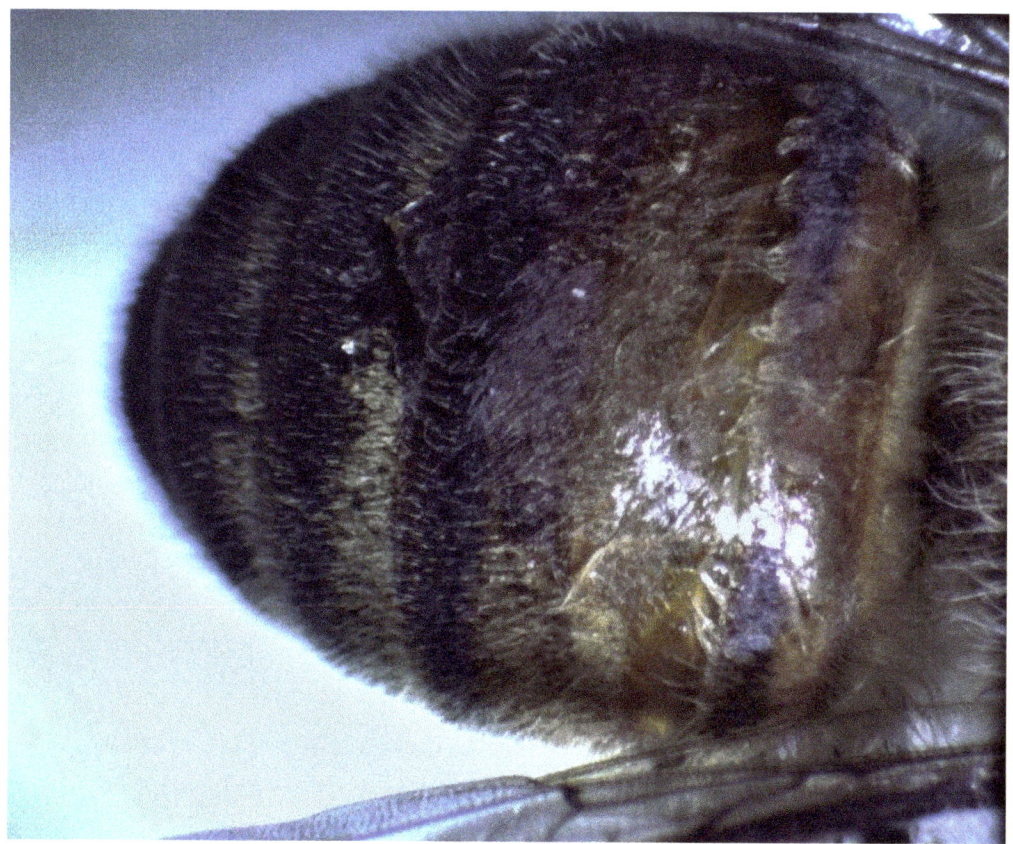
Dorsal abdomen exoskeleton defect

How it should look: The cuticle surface of the exoskeleton should appear smooth and shiny when seen in between and under hair covered areas. Hair should be prominent over the head, thorax, and even portions of the abdomen, eyes, and legs. No breaks or caved-in appearance to the exoskeleton should be seen. Multiple colorations are likely in honey bees. *Make a note to pay attention to the look of the exoskeleton next time you inspect your hive/s.*

Related Pathology: If the exoskeleton is compromised, the following diseases may be involved in either a primary or secondary manner. Brood may be more susceptible to certain diseases because their exoskeleton is not fully developed yet.

Varroa destructor mites enter the body by piercing the exoskeleton in the ventral abdomen and essentially sucking out the fat body of the bee. Additionally, these piercings leave gaping wounds in the exoskeleton, that then allow for the entry of infectious agents, like the dozens of honey bee viruses. These gaps also subject the bee to increased dehydration.

Nutritional deficiencies could prevent the proper carbohydrate and protein balance needed to create the chitin components of a healthy exoskeleton.

Muscle System of the Honey Bee

Form & Function:

The bulk of the skeletal muscles in the honey bee are found in the mid- body section, the thorax. These muscles are used to move the six legs and the four wings attached to the thorax region. These muscles also help to pump air through the honey bee's respiratory system when extra air is needed. They can also aid in vascular circulation of the hemolymph, a bee body fluid like blood. Interestingly, thoracic muscle contraction is required for mammals and humans to breathe every few seconds and aids in cardiovascular circulation.

Honey bee muscles are large and strong, relative to a small individual. These muscles are arranged in two muscle layers longitudinally and vertically. The muscles, anchored to the exoskeleton, work inversely to raise, and lower the wings in flight.

How it should look: Normally (and obviously), you should not be able to directly see honey bee skeletal muscle on a living bee. However, the thorax should have a round, plump appearance with all wings and legs attached to the thorax. Wings and legs should move freely and functionally. *Follow a few bees in your hive at the next inspection, take a few seconds to determine if all the legs and wings are moving properly.* Dissection of dead (but otherwise healthy) honey bees will reveal that the internal thorax will be dominated by its muscle layers. Loss of thoracic muscle tone or wasting is indicative of chronic disease. Injured, deformed, or missing limbs on individual bees could be a typical aging change, but if prevalent throughout the hive, consider chronic parasitic and infectious diseases.

Related Pathology: One downside of having nice, big muscles is that everyone wants some. The meatball like thorax of the honey bee makes them attractive to predators, like birds and Asian giant hornets.

Muscle wasting could be related to: Varroa, honey bee viruses, Parasitic mite syndrome (PMS), and/or nutritional deficiency. Remember it takes protein to build strong muscles for everybody. Loss of locomotion, flight, and/or the ability to use their highly specialized legs is devastating to the individual honey bee and any contribution they make to the colony.

All Things Change

This week the leaves are beginning to change, Fall nectar flow is wrapping up, varroa mite counts are flying up the exponential scale, and my students are taking their first exam on the skin and musculoskeletal systems of the body. With both the good and the bad, seasons change. **Change** is actually the reason for our next study in honey bee anatomy. In continuing our clinical anatomical study of the honey bee and following along with my students' curriculum, the study of the **nervous** and **endocrine** systems are next up in our lineup.

The nervous and endocrine systems are communication systems within the body that control mechanisms designed to maintain *homeostasis* despite the inevitable changes in the surrounding environment. Homeostasis is defined as dynamic mechanisms within the body that detect and respond to deviations in physiological variables from their "set point" values or range. Hive temperature, brood rearing and activity cycles are all examples of variables that need to be monitored and kept within certain ranges to be normal. The goal of homeostasis is to ensure the internal environment of the organism stays the same in an ever-changing environment.

To better understand how these homeostasis maintaining systems play a role in honey bees one must first consider the differences between individual bees and the superorganism colony. While individual honey bees have a nervous system and hormones that are released within their individual bodies, pheromones operate much like endocrine communication for the superorganism. In the superorganism, if the ability to maintain homeostasis breaks down (due to a variety of diseases and external stressors), the colony will eventually collapse.

The Nervous System of the Honey Bee

Form & Function:

The nervous system in animals serves for quick communication within the body. Neurotransmission has been measured at 120m/sec or about 275mph. So, it's fast -think reflex speed. The nervous system operates to adapt and preserve life on a second-to-second basis.

The individual honey bee has a brain and ganglion chain (a high concentration of neurons) located in the head and a nervous cord that transverses through the thorax and abdomen. Like a human spinal cord, the nervous cord of the bee innervates the major muscles and organs of the body. The little bee brain has about a million neurons capable of memory, learning, navigation, and complex social behaviors. The brain also receives and processes sensory perceptions from the outside world. Areas of the brain process visual information coming from the two compound eyes, light information from the ocelli, mechanical and chemical information from the antennae (much like sound, touch, taste, and smell), A special part of the brain, called the mushroom bodies, integrate olfactory information (smell) into memory, not unlike the process of smell and memory integration found in humans.

Short sidebar: Can bees perceive pain? Good question. Personally, I am up to my elbows in this topic right now. I am serving on a committee for the American Veterinary Medical Association (AVMA) working on writing guidelines for Euthanasia and Depopulation for Honey Bees. At this point and in short, I can say this: various research confirms that we know honey bees can detect nociception (receptors for painful stimuli), and integrate it into their nervous system, but no strong research shows if bees perceive what we consider to be pain. That doesn't mean honey bees do not perceive pain, we just do not have the data to show it. It does mean honey bees' welfare is being considered in ways that it hasn't before.

How it should look:

Apart from sensory organs like the eyes and antenna, parts of the nervous system should not be visible in healthy live bees. Changes in behavior are the signs we are looking for to determine normal versus pathological changes. Healthy bees will be about their business, doing the various routine tasks of the hive and moving freely. *Next time you are inspecting your hopefully healthy hive, take some time to just watch all the goings on. Again, learn normal to be able to recognize abnormal.*

Related pathology:

Nothing good happens when disease attacks the nervous system. Infectious diseases and toxicosis are prime suspects with neurological signs. Neurological signs can include aggressiveness, overall quickened jittery or very slow movements throughout the hive, seizures, malaise, paralysis, and sudden death.

The Endocrine System of the Honey Bee

Form & Function:

The endocrine system of animals is made up of glands that secrete chemical signals called hormones into the body. Hormones regulate physiological functions within individual bodies over longer time periods compared to the nervous system. Their effects may last minutes, hours, days, and even control seasonal changes. In mammalian species (including humans) some of our most common and serious diseases are endocrine in nature. Thyroid diseases, a multitude of diseases caused by sex hormone imbalances, Addison's Disease, Cushing's Disease, Diabetes, and pituitary diseases, are all endocrine diseases. Some of which lead to or contribute to death.

Honey bee individuals produce a few hormones with juvenile hormone being the most mentioned in the literature. Juvenile hormone levels regulate division of labor in workers and the differentiation between queens and workers.

There are also exocrine glands in honey bees. Exocrine glands secrete substances outside the body. In honey bees, the hypopharyngeal gland, mandibular glands, various salivary glands, wax glands, and venom glands are all exocrine in nature. However, in honey bees, some of their exocrine glands act in an endocrine way within the superorganism by producing pheromones. Pheromones released by bees act as endocrine signals within and even outside the colony. Examples include: the mandibular gland producing queen mandibular pheromone, the Koschevnikov gland producing alarm pheromone, the Nasonov gland producing a homing signal, drone pheromones attracting drones to congregational areas, and even the brood produces pheromones. All these pheromones are produced in different levels to induce appropriate behaviors at the appropriate time for best colony survival.

How it should look: Much like the nervous system, endocrine organs are not something you can typically examine in live honey bees but changes in behavior may be observed that are normal or abnormal. For example, fanning behavior, getting "tagged" with alarm pheromone after that first sting, and a happily functioning colony with a strong queen presence are all normal demonstrations. Imbalances in appropriate levels of queen, brood, and worker pheromones will lead to dysfunction of the colony.

Related pathology: Any type of queen issue or loss will quickly upset the pheromone balance of a hive. Laying workers is an excellent example of a hive in crisis. Infectious diseases, nutritional diseases and toxicosis can all contribute to the loss of pheromonal balance.

Swarms are one way honey bees reproduce

New Beginnings

Unique to our typical domesticated species, honey bees have multiple ways to "make more bees," depending what perspective one is considering. Honey bee reproduction involves actions at the individual, at the superorganism, and/or at the beekeeping management level/s. The superorganism can reproduce itself by swarming or beekeepers can split colonies. Virgin queens must go on mating flights to interact with multiple drones, a dozen or more, to collect sperm for egg fertilization. Beekeepers can create new queens (and therefore new colonies) by mastering queen rearing and/or queen grafting. Some beekeepers and universities have even utilized artificial insemination (AI) of virgin queens to create new progeny.

Amongst all animals, the reproductive system is the only system that is **not vital** for the survival of the individual. The reproductive system is the last system to receive an organism's resources. An organism will not try to reproduce if it cannot support itself. This is also true of the honey bee colony and why functional reproduction is such an important indicator of colony health. This study will focus on the anatomical and physiological systems of each individual involved, the queen, the drones, and even the worker bees and the effects of their contribution to reproduction within the superorganism.

The Queen's Cabinet

Form & Function:

Much like many animals, queens have paired *ovaries* that ultimately produce eggs (ova). Full development of the ovary only takes place after a queen has been mated. Mature ovaries increase in size up to eight times and occupy much of the abdomen in mature mated queens (1). Eggs develop through a physical migration of portions of the ovary called ovarioles. Queens have more in number and more developed ovarioles compared to other females (workers). These differences are largely controlled by hormones and the differences in food fed to queens (all royal jelly) versus workers.

By themselves, eggs convey half of the queen's genes passed on to her daughters and her sons. Healthy queens are champion egg layers that would put any chicken to shame. During the peak beekeeping season, queens can lay up to two thousand eggs per day, 1.5 million in their lifetime (2). As the egg matures, they move out of the ovarioles to the oviduct and ultimately through the genital chamber to the outside world. Prior to entering the genital chamber, the egg passes by the opening of the *spermatheca*.

The spermatheca is an organ within the queen's abdomen that is capable of storing and preserving collected sperm over her lifetime. The spermatheca is connected to the oviduct within a chamber that contains a muscle and valve capable of allowing sperm from the spermatheca to meet the ova for fertilization... or not. Fertilized eggs result in a diploid individual (two sets of genes, from queen mother and daddy drone). Female honey bees are always diploid. Unfertilized eggs are haploid (no sperm, no genes from any drone, only genes from the queen mother) and result in male honey bees or drones.

Five to ten days after hatching, virgin queens must go on 1-3 mating flights to obtain semen from a dozen or more drones. These matings often result in more sperm than the queen may need to store, and research suggests her reproductive organs ultimately mix and keep 5-7 million sperm to fill her spermatheca while discarding any excess (1).

How it should look

While one cannot examine a live queen's ovaries or spermatheca, one of the best indicators of hive health is the presence of a normally laying queen. Mated queens have plump abdomens compared to virgin queens. Neuroendocrine changes, which occur after mating, change the queen's behavior, and tend to give them away. Queen mothers may move slower than virgins, have a retinue of attendants surrounding them due to their increased queen pheromone, and they may be observed laying an egg into a comb cell. Even if you do not see the queen, the presence of eggs within brood frames indicates that a laying queen has been present in the hive within at least the last three days. Normally eggs (and other stages of brood) may not be present in the late Fall and Winter months but should reappear in the Spring and Summer months. The presence of a healthy queen provides the colony with appropriate amounts of queen pheromone that helps to keep typical processes in the colony going on as normal.

Related Pathology

Queen failure is reported in beekeeping surveys to be one of the leading causes of colony collapse/failure. Lack of brood appropriate to season, hive aggressiveness, low population, generally weak hives, and laying workers are all clinical signs consistent with queen issues. Older queens may fail due to deterioration of their reproductive ability to lay diploid eggs in proper quantities. "Poorly mated" queens may not have collected enough sperm to keep laying diploid eggs and/or provide genetically diverse offspring. Failure of a virgin/newly mated queen to return to her hive, excessive swarming, accidentally killing a queen, and robbing are all common reasons for queen loss.

The Drone's Purpose

Form & Function:

Drones, male bees, are often picked on for seeming to serve little purpose around the colony (other than sleeping, eating, drifting, and having sex maybe once). However, providing genetic diversity is vital to colony survival. Like many other male animals, drones have testicles that produce millions of sperm (spermatozoa). Drones also have seminal vesicles that produce seminal fluid, which makes up much of the semen. A vas deferens carries the sperm from the testicle during ejaculation. Drones do not have a stinger. Instead, they have a large copulatory organ or phallus that is exteriorized during mating. After mating, the phallus breaks off from the drone's abdomen which essentially eviscerates and ultimately kills the drone. The phallus remains within the vagina of the queen and is referred to as the "mating sign." The next suitor drone will typically remove the previous mating sign before mating with the queen himself. This process continues until the freshly mated queen returns to the colony at the completion of her mating flight/s.

Side note * - Since drones lack a stinger, they make good models for handling bees. Use drones to practice holding/marking a queen or for showing honey bees to children. No sting!

How it should look

Mating signs may be seen as newly mated queens return to the hive. Workers typically remove mating signs soon after the queen's return. Drones are typically present in healthy hives in the Spring and Summer months. The presence of some drones is a good sign in hives as healthy hives produce drones only when they have enough resources to support reproductive functions. Sexually mature drones will be drawn via pheromones to drone congregational areas (areas drones hang out and await virgin queens). Also drawn by pheromones, drones will compete to mate with a passing virgin queen. In the late Autumn, any remaining drones are removed from the hive to help save hive resources for the Winter.

Related Pathology

Drone brood due to a longer developmental time can allow for increased propagation of Varroa mites. Many beekeepers may remove or reduce the amount of drone comb in their colonies for this reason. A hive full of drones is trouble and headed for collapse. All eaters and no workers. Lack of a laying queen and the pheromones she produces can cause workers to start to lay haploid eggs, which results in producing only drones. A drone overrun hive is a dead hive buzzing....

The Workers' Yoke

Form & Function:

You may be surprised that workers are included in reproduction, but they play a role as well. Workers are female but are considered sterile in the practical sense. Workers do have ovaries, but their ovaries are small and deactivated during pupal development. Under normal conditions, workers do not lay eggs.

Worker's role in reproduction is incredibly important in the rearing of the young. Workers build combs that house the brood, collect outside resources to provide food resources for the hive, and nurse workers provide for every need of developing brood. Nurses make food to be fed to the larvae through exocrine glands of their bodies. This is somewhat analogous to breast feeding in mammals. We also know that through transgenerational priming, workers play a key role in feeding queens pathogen particles that may enhance immune responses in offspring. Again, this process is somewhat analogous to immunity passed through mother's milk.

How it should look

Workers should dominate the population of the hive and be active in the various roles of the colony at seasonally appropriate times. Brood frames should consist of the highest density of workers, mostly nurses caring for the brood.

Related Pathology

If a colony becomes queen-less for 35+ days, it may become a "laying worker" hive. Due to a pheromone imbalance, workers' ovaries will become active, and workers will start laying eggs. The eggs will often be laid irregularly, multiple eggs in a cell or even outside of the cell. However, because workers are not mated, they are only capable of laying haploid eggs which will become drones. A colony overrun with drones will quickly run out of resources, be overcome by disease and starvation, and ultimately collapse.

Bee "Births"

Before we leave this subject, I thought I would leave you with another analogy. One of the most exciting times in the study of the reproduction system in any animal is the birthing process. While honey bees do not go through a pregnancy and vaginal birth like mammals, honey bees instead have two types of "birth" in bees: when the egg is laid and then the incubation and "hatching" of the brood within the comb. Both processes can be witnessed within a hive and are always fun to watch.

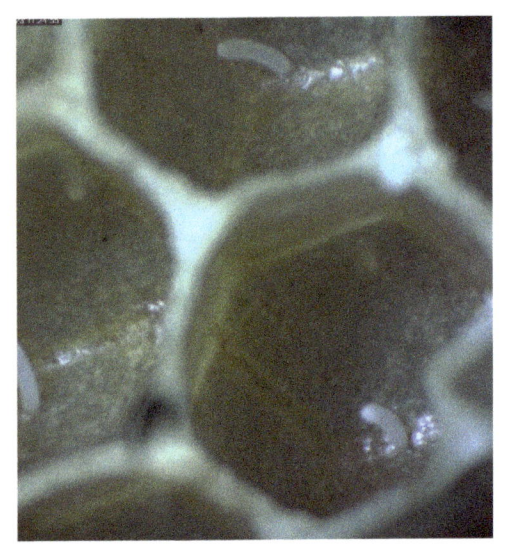

Eggs laid in honey comb

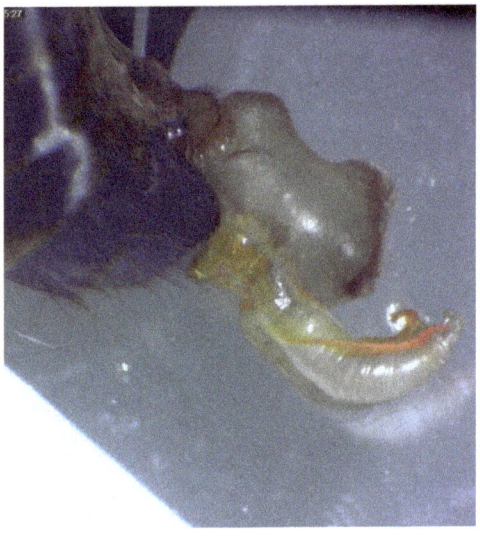

Lateral view of a drone endophallus exteriorized

In with the New and Out with the Old

As we continue to clinically explore the systems of the honey bee, we will be spelunking through the caves of the digestive tract and excretory tubules of the bee's body. Not surprisingly the superorganism displays a version of these functions throughout the colony. Meanwhile, my anatomy students have just finished their exam on these same subjects within the human body and achieved record scores! Even after teaching the subject for years, I am still amazed at the remarkable anatomical similarities found throughout the animal kingdom, from the human being to the honey bee, aside from "slight" functional design modifications, of course.

The Digestive System

Form & Function

Within any animal, a digestive system acquires solid and liquid resources from the outside environment. The digestive system has the ability to temporarily store, but ultimately break down and sort these resources into usable nutrients the organism requires for survival. Any excess, useless, or waste products must also be eliminated by this system. The honey bee's digestive tract is not unlike that of many other animals in its complex function and specific design.

The digestive tract begins with the tongue and the oral cavity of the bee. The tongue is useful in collecting and transferring nectar, honeydew, water, and honey for individual use or for sharing with others. The relatively long tongue or proboscis is complex in design and functions to probe nectar sources, provide suction, and gather desired liquids.

The gut continues from the oral cavity and is divided into different functional and anatomical portions. These portions are synonymous with other animal and human digestive tract divisions. Like many other animals, honey bees have an esophagus which simply acts as a tube that transports materials "from point A to B" through the thorax to the remainder of the gut located in the abdomen. Smooth muscle lines the entire tract and helps to move the ingested material from esophagus to anus.

At this point, the gut demonstrates a honey bee modification called the honey stomach or crop. Here the honey bee can store and carry nectar, honey, or water for transport. No breakdown, digestion or absorption of the food products occur here. The honey stomach can greatly expand to accommodate a large volume and/or regurgitate its contents to be given up in trophallaxis. The crop is like a grocery bag that can be emptied when the honey bee is ready to share what she has accumulated with the rest of the colony.

The proventriculus is next in line and acts as a regulatory valve between the honey stomach and the true stomach or ventriculus. Once materials enter the ventriculus, they are potential absorbable nutrients for that individual honey bee's needs. Enzymes designed to break down carbohydrates and proteins to smaller absorbable products are all released in this true stomach. Much like mammalian stomachs, the ventriculus is composed of a twisted shape and folds designed to allow the ventriculus to create voluminous surface area for food stuffs to have plenty of opportunities to mix with the enzymes secreted there.

Again, much like other animals, the next stop after the stomach in the digestive tract is the intestine. The first part of the honey bee's intestine, or anterior intestine, is narrower than the second portion of the intestine, or the posterior intestine or colon. The small intestine in animals is much like the bee's anterior intestine and the large intestine or colon is synonymous with the honey bee's posterior intestine or colon. Functions are also remarkably similar throughout the animal kingdom. Absorption of most nutrients into the bee's body is complete in the anterior intestine. The jobs of the posterior intestine or colon are to consolidate waste and reabsorb precious water. Those basic functions are the same for you and your pet cat.

Finally, the digestive tract comes to an end at the rectum and the anus. The rectum is utilized for the storage of fecal material and empties into the anus into the same area of the sting chamber. The anus serves to open or close the rectum. Honey bees do not defecate in the hive. Therefore, in the winter the rectum may expand to occupy the greater part of the abdomen until the honey bee can go on a cleansing flight.

Other important components of the proper functioning in the gut includes the microflora inhabitants of the gut. These microflorae are bacterial and fungal symbiotic beings that aid in digestion and immunity of the honey bee. Truly both in honey bee science and human enterology we are learning more and more about the importance of these microorganisms in the health of our gut and immune systems.

Food processing and storage of honey (carbohydrate source) and pollen (proteins, lipids, and micronutrients) also occur in the colony as in the individual body. If you look at it from a superorganism perspective, various nutrient stores in the hive are much like fat, carbohydrate and protein stores found throughout the body in other animals. Normal microflora within the colony are also vital to the processing and storage of food within the hive. Nurse bees are able to produce food for the young through various glands located in and around their head. While mammals take in nutrients to make milk in glands for their babies, nurse bees can do the same to make brood food and royal jelly to feed to their little siblings.

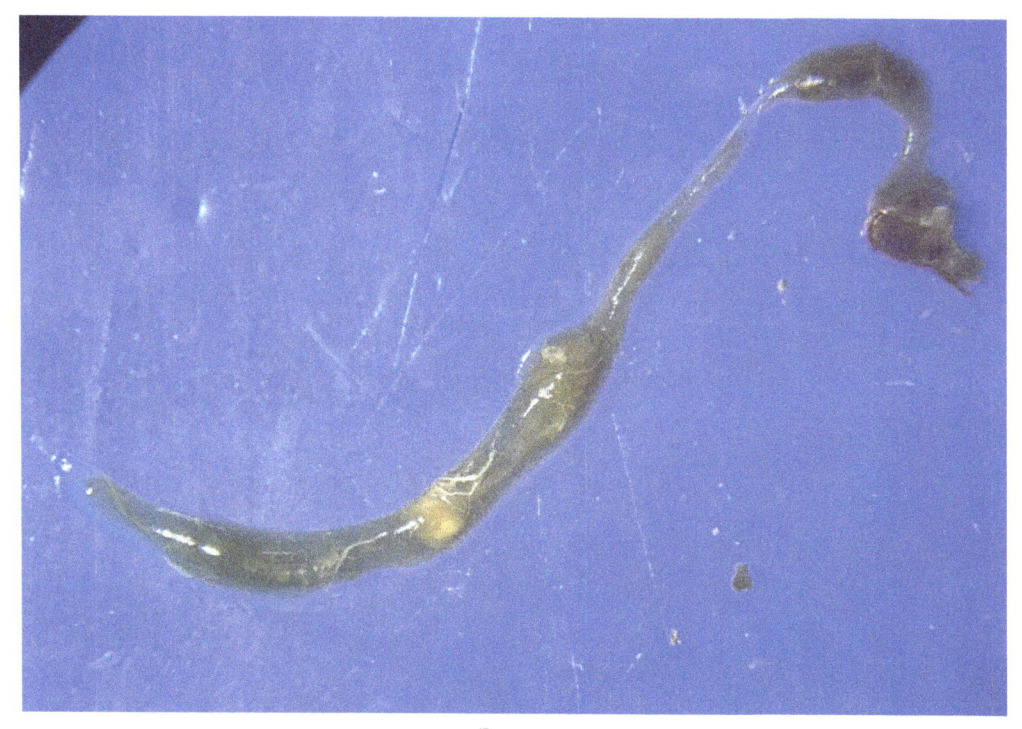

Bee gut

Excretory system

Form & Function

Excretory comes from the word "excrete," meaning to actively remove a substance from a given position. Any animal's excretory system is involved in the removal of excess, useless, or toxic products that are acquired through ingestion or other exterior exposures, and internal sources from by-products produced by the organism's own metabolism. Working with the digestive system, the excretory system helps to cleanse the body of unwanted substances. In mammals, the kidney's filtering of the blood is the crux of excretory organs function. In the honey bee, the Malpighian tubes are considered to be kidney-like organs. Much like the kidneys that filter 20-25% of the cardiac output in a mammal, the Malpighian tubes filter the openly flowing hemolymph (or bee blood), remove waste products, and balance electrolytes and water. Kidney and Malpighian tubes anatomy is similar in that the tissue is primarily in a convoluted tubule form. However, the honey bee does not have a urinary tract or bladder but instead the Malpighian tubes empty into the digestive tract at the junction of the ventriculus and intestine. I suppose this is an efficient means to share the way to the exit. This design is somewhat like the birds, as bee droppings are a combination of bee "poo" and "pee."

From a colony perspective, the fastidious behavior of the hive inhabitants provides an excretory function for the superorganism as a whole. Examples include, undertaking bees removing the dead, sick bees flying away from the hive and cleaner and guard bees constantly removing debris or invaders from the hive. Since honey bees also do not defecate within the colony this also reduces waste build up.

A forager's gut is complex

Disease Implications

Dysentery - Honey bees can get the equivalent of diarrhea. Spotty, long smears of feces all over the hive and/or hive entrance is an indicator of dysentery. Dysentery can be caused by infectious agents like *Nosema apis*, poor nutrition or long-term confinement/lack of ability to take cleansing flights.

An external hive exam may be all you need to notice a digestive issue. Watch for runny bee droppings all over your hive/ hive entrance area. Be sure to observe the difference between normal cleansing flight activity and dysentery.

Toxins - The digestive systems and excretory systems of honey bees and other animals face exposure to whatever substances may be found in the food and water they collect. These two systems specialize in maintaining the homeostasis of the body with individual honey bees and the superorganism by sorting out wastes and eliminating toxic substances from the body. Various pesticides, heavy metals, and even some natural toxins from plants can challenge the digestive and excretory system of the bee. Most studies point to a cumulative effect of toxin exposures as an added stress that can lead to loss of homeostasis and what beekeepers call colony collapse or loss.

While most toxin exposure is chronic and can be difficult to clinically diagnose, acute cases may show substantial amounts of dead and dying bees at the entrance of a hive, along with seizure-like behavior.

Poor nutrition is another disease that can have a profound effect on the digestive system. Without the proper amounts and balance of food stuffs, the gut cannot function to absorb and provide the body with required nutrients. The microflora of the gut will also suffer without proper nutrition and immunity will decline due to the loss of the microflora and proteins necessary to equip the immune system.

Be sure you have a seasonal system to assess if your bees have the proper nutrition available for them and if not, how to supplement them. Be sure to thoughtfully check out the local forage in an area before placing honey bees there. Be sure your honey bees have enough honey to overwinter. This may be 40-80 lbs. of honey per colony depending on your local climate. You may have to feed your bees in any season as needed. Know when and what to feed your bees: honey syrup 1:1 or 2:1, fondant, dry sugar, pollen supplement or substitute. Staying connected with hive inspections and weather assessments is critical to informing your decisions.

Even in an ever-challenging environment, the digestive system and the excretory systems of the honey bee are beautifully equipped to bring in the new and remove the old. If you have ever parked your car outside a bee yard in late Winter, perhaps you have enjoyed the fruits of cleansing flights all over your vehicle. While it can be a little tricky to scrub all those little yellowish speckles off your paint, it brings to mind the hope and promise of Spring's nearing arrival.

Take a look at the systems of your hive.

Bee Beats and Breaths

As I am writing this, my anatomy students are plodding across the finish line of finals, and I am literally proctoring their last exam focusing on the cardiovascular and respiratory systems of the body. Christmas décor abounds and I just put some sugar on my bees.

Like the other systems we have reviewed, the cardiovascular and respiratory systems of the honey bee are much like other animals in form and function with modifications designed to accommodate a busy, small body. Unlike many of the specialized systems we have previously discussed, portions of the cardiovascular system and respiratory systems are present in all parts of the bee's body head, legs, wings, thorax, and abdomen.

The Cardiovascular System

Form & Function:

One of the benefits of being relatively small is the ability to get by with an "open" circulatory and respiratory system. There is no need for an extensive vessel system to carry blood meters away and back throughout the body. Bee "blood" called hemolymph, can simply bathe and flush through the tissues for nutrient and waste product diffusion exchanges. Hemolymph is an appropriate name for this honey bee circulatory fluid, as "hemo" means blood and lymph refers to lymphatic fluid. In larger animals, lymphatic fluid mirrors the venous circulation and picks up any extra fluid that may leak out of the blood stream bringing it back to the heart. Honey bees are able to combine the form and function of two circulatory fluids into one.

Hemolymph is a clear to straw colored fluid, not red like our blood and many other animals, because bee blood does not need hemoglobin to help carry oxygen. Hemoglobin, which reflects reddish pigments, is a protein complex utilized in bigger animals and humans to help carry copious amounts of oxygen through the long haul of our systems. This is not necessary in an efficient little bee. Hemolymph, also somewhat like other animals' blood, contains blood cells or hemocytes and protein that circulate through the body and are tasked with various immune functions.

The honey bee's heart is a relatively long tube that runs through the dorsal thorax and abdomen. It has openings that are valve like structures called ostia that help blood move in one direction of circulation. The heart serves to pump hemolymph from the abdomen to supply the head. Much like other animals the largest vessel in the bee body is the aorta. This large vessel carries hemolymph from the heart to the head to ensure perfusion of the brain and sensory organs found there. The bee blood then trickles back through the bee's tissues from the head, through the thorax, abdomen and back to the heart again. There are even accessory pulsating organs found in the bee's extremities to keep circulation moving in the legs.

How it Should Look

Clearly you cannot and should not be able to directly examine the cardiovascular system of a healthy live honey bee. However, you can get an idea of their hydration status which is related to their cardiovascular success. *Look for: The cuticle of the honey bee exoskeleton should look waxy and shiny. Their bodies should be full and not pitted or sunken. Open brood should be juicy and moist, covered with brood food.*

The Respiratory System

Form & Function:

Do bees breathe? Yes, but not exactly like we do. Honey bees have multiple openings throughout their body called spiracles where air can enter and exit the body. Remember that the bee's body is covered with protective hairs that also guard the openings of the spiracles, so hopefully nothing but air can get in. We have similar hairs that protect our respiratory openings...ever take a look up your nostrils?...whoa, lots of hair! The spiracles' openings also have valves that can close to keep unwanted things out. In humans, our larynx does that for us.

The spiracles lead to multiple tracheae, or air tubes that branch into small tubes which ultimately lead to tissues and organs to supply them with oxygen and remove carbon dioxide. There are also air sacs in the bees' body that can act as a reserve for air and pressure. Built up air pressure can help with circulation of airflow and hemolymph. Much like what taking a deep breath can do for you. People and most larger animals only have one trachea leading to a large bronchial tree of thousands of air tubes leading to clusters of air sacs or alveoli. Again, scale is the difference here.

How it Should Look

After high metabolic activity, honey bees can be seen "pumping" their abdomens to increase circulation and airflow within their bodies. Just like we can use more muscles to increase our respiratory effort, so may the bee. It is like breathing hard after exercise. *Look for foragers landing on the top of the hive pumping their abdomens... they may just be taking a "breather" after a long run for nectar or pollen.*

Superorganism Systems

From the superorganism perspective, the collective can be perceived as performing cardiovascular and respiratory-like functions. In- hive worker bees act as a transport system to move moisture, nutrients and protective propolis around the hive as needed by the entire superorganism. The whole hive can breathe by circulating airflow as observed with fanning.

Potential Pathology

Keep in mind that any product that could clog the spiracles of a honey bee could potentially lead to "drowning."

Visible, pronounced tracheae in developing larvae can be a sign of European Foulbrood.

Tracheal mites or *Acarapis woodi* are mites that infest the trachea of honey bees. They caused problems in US honey bee colonies in the 1980's but with the arrival of Varroa mites, tracheal mites faded in pathological comparison. These mites are not visible to the naked eye but must be viewed with the help of a light microscope or even electron microscope. The good news is these mites are controlled with the use of miticides for Varroa control.

So do all topics on keeping bees healthy lead back to "do regular mite counts with alcohol washes and treat for Varroa" ...well, yes.

School's out!

As I am finishing this writing, my last student just handed in the last final of the year that I am giving! Time for a break! No worries, no quiz for you, but I hope you have enjoyed this review of honey bee anatomy from a comparative and clinical perspective, as well as a trip along with my students' educational endeavors. Remember to recognize abnormal you must know what normal is. Applying this practice will make you an observant and more effective beekeeper in the bee yard.

Frame Four - Honey Bees Behavior

Human and animal behavior play a special role in the study of biology. How an animal reacts to the environment or not can impact its health and wellbeing. Again, differentiating between normal and abnormal behaviors is paramount to a beekeeper's ability to interact with their wards safely and successfully. This frame features commentary on a few common keeper-bee interactions.

All the tasks of the hive require healthy behaviors

Human-Bee Bond?

Mutually beneficial relationships have existed between humans and animals for thousands of years, but in the1990's the recognition and importance of the **human-animal bond** began to permeate veterinary medicine and our society in general. As dogs and cats moved from the backyard or the barn to the house, our animal's status changed from pets to family members. Many pet "parents" today find it perfectly normal to share much of their beds with their beloved canine, feline and/or other furry creatures.

We and our bees

While attending veterinary school at Ohio State University in the mid to late 90's, I was one of the founding student officers in our Human Animal Bond Club. At that time, we did not fully understand the meaning of our club's purpose, we just liked bringing animals around to visit folks in nursing homes. Since then, I have watched the human-animal bond relationship evolve from a near front row seat. Volumes have been written on the subject in the past 30 years. Laws have changed. How we handle animals in scientific research, in veterinary medicine, and in agriculture has changed. How we acknowledge and treat animal pain has changed. How we value animals in society has changed. The human-animal bond has become its own subject of study (just check Google scholar), and has meaningful, beneficial, and economical applications for both humans and our animals.

During one of the first bee vet medicine disease courses I took, another veterinarian asked if bees recognize their keeper. "No.. (with a laugh)", was the response from the instructor. However, over time, my research and observations with honey bees and beekeepers are leading me to another conclusion.

I am not saying that your bees are going to love you like a devoted, slobbering Labrador Retriever. They will not. But let us consider what science, both biological and social tell us. I will also throw in some anecdotal evidence that I believe most keepers can relate to.

Behavioral and Physiological Science:

There are many reasons for aggressive or defensive behavior in bees: genetic predisposition, extreme weather, queenlessness/pheromone imbalance, underlying diseases, and lack of reserves/hunger. Notice these reasons are not unlike the major causes of defensive action in all animals. One factor that induces defensive behavior that you can control every time you interact with your hive is your behavior. Honey bees, like most animals, recognize and communicate threats and non- threats. They will change their behavior toward you based on their assessment. Anecdotally, after shadowing dozens of beekeepers, I have observed that "rougher" beekeepers tend to have "crankier" bees.

Honey bees' brains have incredible capacity for memory, and they are capable of passing this memory on to the hive collectively. Honey bees have keen visual, mechanical, and chemical senses. If you visit them on a regular basis, it is likely they know what you look like, smell like, and whether you and your behavior are perceived as a threat or not.

Social interactions:

Many backyard beekeepers may view their hives like pets. One's hives may even make your Christmas card picture along with your kids, your dog, your cat, and your spouse! However, the human-animal bond does not just apply to pet owners. Anyone who has spent some time with a family farmer knows that most love and care for their animals. Growing up in dairy farm land and in my early vet days, I can remember watching dairymen calling in their cows from the field for milking. A cow's ear tag may have said #49, but to the dairymen, she was "Susie". He would call her by name, she knew his voice, and he helped direct her to her stall. To me, they looked like a bunch of black and white Holsteins, but to the farmer these animals were his life and livelihood. Granted the dairyman knew if Susie's production dropped, she would have to "go down the road", but he still stewarded an empathy for her life's contribution to his family. In commercial bee yards, I have observed a similarity amongst beekeepers with bees much like a dairyman caring for his cows. I have heard these bee farmers lament and struggle with the choice of moving their bees and disturbing their peaceful "porch sitting" evening. Many bee farmers worry about their bees' health and stress, understanding the relationship that their bees' wellbeing has on their own.

We and our bees

Other Bond Benefits:

Included in the definition of the human-animal bond are the mutual benefits of the bond between the animal and human, including emotional, psychological, and physical aspects. The environment also plays a role in the interaction of this relationship.

On the bee's side, a good keeper will provide proper shelter, nutrition, health and preventative care. Unfortunately, without knowledgeable, involved keeper intervention many *Apis mellifera* colonies will perish within a few years due to pests, disease, exposure and/or poor nutrition.

On the human-side consider everything bees do for us! Like many agricultural animals, we may feed them but ultimately, they feed us. Bees provide honey, wax and other hive products, **and** pollinate a substantial number of crops that humans and other animals eat. Beekeeping itself also provides a means of physical activity. Our benefits from honey bees go much further than just the physical benefits. Beekeeping is acknowledged as an activity that can improve mental and emotional health, one of the biggest public health issues we face. Many beekeepers have expressed how their hives gave them something positive and "safe" to do during COVID-19 lock downs.

Ways to Improve Your Bond:

Know your bees. Keep records. Learn how to do regular, gentle inspections. Move smoothly, at "bee speed". Try not to crush too many bees or bang on the hive. If you are afraid to inspect your bees on a regular basis for whatever reason, get some help. Your developing knowledge and technique will keep them healthier and both of you "happier".

Try less smoke. Imagine if every time you met someone; they thought their house was burning down. Yes, there are times and places for smoke. I am not saying throw your smoker away. Certainly, commercial settings or dealing with Africanized bees are a different story. But a smoker is essentially what we call, in veterinary medicine, a restraint device.

A restraint device is a tool or technique that veterinary professionals use to control an animal's movements for the safety of the animal, the animal owner, and the veterinary professional. However, if too little or too much restraint is used, specific to the situation, control and safety may be lost. An example of a restraint device would be a dog muzzle or a cat bag. Most veterinary professionals will start with the "less is more" approach, especially with felines. Most of the time this is the best approach for most cats. However, a few cats may still end up jumping to the ceiling, no matter what you have tried. I have found that *Western/European* honey bees to be somewhat like cats, in that less smoke is more for most hives, on many days. Keep your smoker lit, but use it judiciously, only when needed.

Food for thought:

I am writing this in August in Western Pennsylvania, where we are experiencing an extended dearth, exacerbated by severe drought conditions which we have not experienced for years. One recent evening while sitting on the back deck with my husband, I had bees hovering and crawling all over me, but not my husband. "They know you," he said with a smile. I think he is probably right. I have been in so many hives I probably "smell" like a queen pheromone cocktail mixed with sugar syrup. Every evening my bees follow me around like a puppy, as I water my plants, hoping to get a sip of precious water. They just want to be fed, right? ... But is this not a major way we form bonds with most animals? We feed them. Nothing wins over a cat like their favorite treat. My horses come running into the barn after I call them, yes ...**and** after I dump grain in their buckets. The very first human relationships with canines developed thousands of years ago around our campfires where food scraps were easier to come by than typical prey. Are our bees really that different?

The human-bee bond. A bond with tens of thousands of stinging insects? I believe we and our bees absolutely meet the definition of this important relationship... but these babies are still staying outside in the yard for me.

How to Stop Swarming

With a title like that, I know I got your attention. Essentially, there **is** one sure way to **stop** swarming. Kill your bees. A dead colony will not swarm. If you prefer to keep your colonies alive, read on for a biological, health based and practical take on how to manage this topic of much discussion.

Personally, I believe swarming is an overdone topic in the beekeeping literature. Many have opined on the subject to the point of ad nauseum per my appetite. So up until this point I have resisted writing about it. However, people have asked me to write on the subject...so here goes.

I have observed that the honey bee swarming phenomena involves and interacts within three major aspects of beekeeping: the biology of the bee, the sector of the industry, and the psychology of the beekeeper.

Our interactions with swarms

The Psychology

There is no denying that swarming is a fascination amongst the general public and many beekeepers, especially newer beekeepers. Many want to know the "secret" to stopping it, controlling it, like a quest for the Holy Grail. All types of gadgets, equipment, sprays, and folks dressed in bee beards in precarious, peculiar places are devoted to the capture of this phenomenon. Beekeepers have favorite swarm stories to tell and to one up the next guy. Is it the excitement of the chase? The wonder of Nature's workings? Or the "cool" (and somewhat narcissistic) reputation of being a bee wrangler?

Why do some hunger for more information on this age-old subject? I suspect it is analogous to weight loss. Everyone is looking for the next magical cure, gadget, plan, or pill. Everyone also already knows what it really takes to lose weight is obvious math (do not eat more calories than you burn throughout your entire life) but that's consistent devotion, hard, and boring. We also already know how to *manage* swarming, but it takes a consistent plan, and it is still going to fail sometimes. But of course, this approach lacks ease and pizzazz.

I have also observed that most beekeepers, if they make it past a couple of years, go through what I call the progressive Stages of Swarm Management: Stage 1) Amazement, Stage 2) Annoyance, Stage 3) Acceptance. Stage 1, Amazement, dominates in newer beekeepers. That first swarm you successfully caught or that weird local circumstance- Amazing! But after a few annoying phone calls at exactly the wrong time from unhappy neighbors and/or loss of your favorite queen.... Stage 2, Annoyance enters the arena of the mind.

Recently I had the opportunity to take a less than 24-hour horseback riding/camping trip into the "Wilds" of Pennsylvania with a few family members. Which means I dared to leave town overnight. That same day a swarm occurred from a split I had made the previous day to *prevent* swarming at my campus's apiary site. But the moved mated queen did not like her new abode, so she moved about ½ a mile away, right in front of the main science building on campus on a small ornamental tree lining the main walkway. Yes, in front of the building in the CENTER of campus with HUGE windows. Once I got back into cell phone range that evening, my phone was blowing up with emails, texts, and calls from everyone and someone's mother about the ominous swarm on campus. I arrived the next day to see yellow caution tape and a crowd of nervous on lookers wondering why I had not commissioned a jet to arrive sooner. The modest size swarm was resting peacefully on a lower branch of the small tree just within reach. My students and I easily caught the swarm, I gave the insurrectionist queen and her colony to a gracious beekeeping neighbor, and I have provided repetitive educational talks and pamphlets about swarms to the concerned crowd. Yet I continued to receive paranoid calls to come and get the swarm again over the next several days, because clearly, I missed the swarm. I drove back to the area for several days to witness a finger full of bees clinging to the branch's residual queen pheromone, caution tape still intact. I prayed for rain. It still did not matter. I was officially revisited by Stage 2.

Stage 3, Acceptance comes along with time, patience, and perseverance. Maybe old age and just being tired. It brings peace to accept that you cannot always control everything despite your best efforts within the best practices. Of course, portions of all stages will still be represented in the emotional and psychological state of the beekeeper. However, with experience and time most beekeepers mature to Stage 3 dominating their overall outlook of swarming with various inklings of Stage 1 and 2 still within their minds.

Industry Management Caveats

From what I have observed, swarming concerns are found within the common vernacular of backyard to side-liner beekeepers. Commercial/migratory beekeepers, those that make a living doing this and manage most colonies in this country, have bigger issues to deal with and innately manage swarming within their typical SOP. Commercial beekeepers manage herds of honey bees by the yard or truckload. Each group has a working purpose, whether pollination, making nucs, packages or queens. These beekeepers keep their hives relatively small, busy, and moving. Inspections are done by the yard. Inspections are done quickly and decisively, with queen replacements in their back pockets. Their colonies are typically not big, fat, tall backyard hives with nothing better to do than swarm. If they do happen to park their hives in a clover field to "rest" for the summer, they mostly let them be... bees, and replace queens if/as needed. Economically, it is simply not beneficial to them to chase after possible swarms. They live in Stage 3.

For the rest of us, swarm management is not about stopping swarming. Anyone telling you that completely stopping swarming is possible is either an oblivious beekeeper, a gas lighter, trying to sell you something, or all the above. As a vet, if there was a way to "neuter" honey bees, I would have found one. There is not. Swarm management is about awareness of the honey bee's biology, what your goals are, and how you can intervene to provide stewardship to this natural process.

The Biology and Health Considerations

First if you are not in your "queen-right" hive/s every 7 days within the beekeeping season – you simply do not know what is happening inside that hive. If you only do inspections on your hives once a month (or less) you **will** miss many swarms. That is not my opinion, that's biology. Ignorance is bliss is not a strategy based on reality here. Those that believe/claim that their hives "never" swarm, live in this happy and clueless place.

Swarming is a natural process that occurs in healthy, often vibrant, honey bee colonies as part of their reproduction cycle. Swarming (but not absconding) indicates that the hive is healthy and has enough reserves to reproduce. It has some benefits to the bee and beekeeper and some downsides. It provides a brood break for the stationary colony that is helpful in reducing or slowing down various diseases that affect the brood, including our chief nemesis, *Varroa destructor* and the many viruses it vectors. It allows for re-queening, through a returning virgin or an introduced mated queen. Re-queening is often a remedy for getting things back on track within a hive, whether it be disease recovery, poor brood patterns/faltering old queen, or colony behavioral problems. The swarming nature of the honey bee, in part, allows us to utilize their system to create new hives to replace those that are lost within the current disease and environmental challenges our bees face.

One should also consider your purpose with your honey bees. Do you just want a few hives? Do you want to make a lot of splits? Do you want to make more honey or nucs? What equipment do you have available? Do you really like that queen or not? How much effort you put into controlling swarms may depend on what you are trying to do. There are various actions we can do to slow/manage swarming if that is your desired effect.

Things that slow swarming down:

1. Adequate space before they start to swarm. Whatever that means to different bees may vary. Drawn comb is better space.

2. Clipping mated queen's wings, so she cannot fly away... I have had a clipped queen swarm onto the ground ten yards away from her original hive. One must appreciate her grit.

3. Reversing boxes- even deeps. Bees like to go up. Reminding them they have space in the basement can sometimes slow swarming.

4. Knocking off empty queen **cups** – might give you another week. If it is a queen cell/swarm cell it is too late to stop the swarming behavior, but you may still be able to find the queen mother and move her. If it is capped, it is too late. She has gone and now you are left to decide what you would like to do with her pupating daughters.

If you want only the original hive, knock off all queen cells but two or you can divide up the hive and try to make multiple nucs per couple of cells. If you do not know the difference between a queen cup and swarm cells, do some reading on the topics asap.

5. Splitting. Move the queen mother if you can...to another yard if you can, with more honey, less brood frames. Re-queen or be sure to leave eggs in the original colony and recheck in a week for the appropriate number of queen cells/emergency cells.

6. Caging the queen. Personally, I have not done this so I cannot speak to it much. But if the queen is in jail she cannot leave, it provides a brood break, and a population hold. It also requires intense management timelines. You must go back in and release her at some point and the process could damage the queen and/or negatively change the pheromone balance within the colony.

7. Some races/genetic lines may be less likely to swarm or may be more likely to swarm. You can experiment with it as you wish. I would recommend buying an extra box of tissues for when you watch your designer queen sail fifty feet up into the forest.

8. Telling the bees you are leaving so they do not swarm. Ok so there is no scientific proof for this, but it is an endearing adage. For me personally, if correlation equaled causation, the data would be compelling. Much "science" has been proclaimed with less.

Things that do not help slow swarming (as much as one may hope):

1. Undrawn comb for "space".

2. July-November. Swarm season is over! Nope, not really.

3. Fencing.

New queen emerging

Reasons to try to prevent swarming:

There are reasons beekeepers should consider making an effort to manage swarms.

1. Spreading feral hives into the environment and the problems therein.

2. Increasing competition with other native pollinators.

3. Increasing disease spread.

4. Unhappy neighbors.

5. Public health safety concerns.

6. Phone calls, texts, and emails at the most inopportune time.

7. Ultimate death of the swarmed, now feral colony.

8. Loss of control of queen/genetics in the hive.

9. Risk of failure to re-queen.

Reasons to not feel bad when they swarm:

1. Keeps hive size more manageable. Any more than 5-6 boxes are taller than me anyway.

2. Brood break.

3. Fresh virgin queen, a chance to change age, performance, disease, attitude.

4. It is a positive indication of hive health.

5. Slow brood diseases.

6. Keeps honey in one hive. Instead of splits. You ran out of equipment anyway.

7. Fun playing with queen cells/swarm cells.

8. Chance to try/introduce a new queen.

9. You are human and need a life outside the bee yard, too.

The Last "Take Away"

Hopefully, you have noticed that I have tried to supply a summary of practical advice on swarm management along with some beekeeping humor to keep it real. My best overall advice is (and I have not really heard this much in the literature): Have a seasonal swarm management plan fit for your purpose and expect the bees to not always read the books, ESPECIALLY AS FALL APPROACHES. I hear a lot of talk about swarm season in the Spring months and while it is true that swarming more commonly happens in the Spring, it can also happen throughout the summer and fall, especially if you have good weather conditions and healthy bees. The problem with this time of the year is with less time to requeen and recover, Fall is the time when swarming can be "deadly" to colonies. Spring has time and resource forgiveness for swarming. Fall does not. I would suggest that this is the most crucial time to make a plan for the possibility of swarming.

Do you have a late source of queen replacements? Are you going to combine hives? What if it gets too cold to inspect? Awareness of how to prevent or take your "losses" in the Fall may be a good strategy for making it through the Winter and flourishing the next Spring.

The Conundrum of Feral Honey Bees

I live out in the middle of the woods, in what most people would think of as the "boonies." I love the gift of nature so much that I became a biologist, a veterinarian, and a self-proclaimed tree herder. I tend to enjoy "natural" remedies and prefer to use medications very judiciously for myself, my family, and my patients. My husband and I built much of the home we live in. I like to forage through the woods, and he hunts. It is all very romantic hunter-gatherer stuff and I love it. Especially when I can come back to a house with running warm water, electricity, and sometimes decent internet. Above all I am a realist. Truth be told, Mother Nature can be a real bitch. Fortunately, we have been given gifts of knowledge to manage the throes of this lovely lady.

Feral is a word we may use in veterinary medicine to describe various animals, like cats, hogs, and dogs. Even horses have gone wild from time to time. Meriam-Webster definitions for "feral" include references to "wild beasts," or "not domesticated", and/or "having escaped from domestication"…something that **became** wild. These definitions imply that we are referring to an animal that was intended to be domesticated. True wild animals have never been domesticated or are not described as feral. Origin of the word, "feral" use dates to the early 1600's, which I find interesting since (domesticated) honey bees were being first introduced into the American colonies at that time (1). Agriculture and the domestication of animals dates back at least 10,000 years, **so we have been here a long time.** Honey bee domestication is thought to not be far behind canines and ruminants at about 9,000 years ago (2,3).

"Sylvatic" (meaning woods) is another term we use often in biology and veterinary medicine to describe things (often diseases) that occur, affect, or are transmitted by wild animals (4). One of our many goals in veterinary medicine is to keep as much of the "wilds of the woods" (i.e., disease) away from our domestic animals **and** human populations as possible.

One benefit domestic animals have enjoyed by teaming up with humans is a great increase in population and diversity. What was once a small population of the African/Middle Eastern wildcat, *Felis sylvestris*, which means "cat of the woods" (and the inspiration for Tweedy Bird's Looney Tune friend) became our domestic cat, *Felis catus or domesticus*, one of the most populous animals on Earth (5).

Along with cattle, pigs, sheep, goats, horses, donkeys and dogs, honey bees have also followed this trend. Tens of millions of hives are managed all over the globe with the US's 2.7 million colonies barely making the top ten of countries reporting colony numbers. India boasts about 12.2 million hives. China and Turkey rank second and third with 9.2 and 8.1 million hives, respectively (6, 7).

Why do our domestic animals do so well compared to feral or actual wild animals? Because we care for them. We feed them. We shelter them. We prevent and treat them for diseases. Admittedly, there have been and are poor management practices that can be a detriment to our animals. I believe evaluating and reevaluating methods from multiple management perspectives to combine best practices is well worth pursuing.

Domestic honey bees naturally swarm, this is when colonies could become feral. It is like my house cat running out the door. If I do not catch them, they will be gone. Unfortunately, in this situation my bees and cat will be subject to a higher risk of disease, injury, shelter, and nutritional issues outside of my care.

As a veterinarian, I have recovered many feral animals, mostly cats. They are often infested with parasites, fleas, ticks, and intestinal worms. Some have bacterial and viral diseases. Many have injuries, stunted growth and weight, and skin conditions. Cats (and some other feral species) are particularly good at "surviving" in the wild, but I can tell you they do it with suffering. A few targeted treatments, healthy food, and a safe place can turn these very sick patients around in a week.

Feral animal populations create threats for both domestic and wild animals and sometimes human populations by acting as "reservoirs" for various diseases. Wild dogs and wild hogs pose a threat as a reservoir of diseases and also a safety threat to other animals and humans. For these reasons in public health, we try to create physical and/or medical boundaries to separate wild and domestic animals. Feral animals often walk this line and may vector diseases from one realm to another. Feral and wild animal reservoirs are one reason we may never be able to eliminate a particular disease from an area. Rabies is a good example of this. Rabies has the highest mortality rate of any virus, nearly 100%. Luckily, Rabies is not contagious through airborne means and many wild or feral animals that do contract it die quickly before further transmission is possible. In many states Rabies remains endemic in our wild and feral animal populations but limited in domestic animal **and** human populations due to education, avoidance, and highly effective vaccines.

Pathological relationship outcomes between various disease agents and animal populations are not at all equal. Outcomes often depend on the type of disease agent, ex. virus vs. parasite, the virulence and mortality rate of a disease, and the overall health of the host. Terms like resistance, adaptation, tolerance, immunity, parasitic relationships, symbiotic relationships, subclinical infections/infestations are all thrown around in the literature. They all have different meanings and applications; they are not the same.

For example, domestic grazing animals are still exposed to an environment contaminated with intestinal parasites, (yes, the dirt is loaded with a wide variety of parasite eggs and larvae that animals and humans can ingest). Parasites do not typically want to kill their host. Since eliminating them in the environment is an impossibility, we satisfy this threat by strategic deworming programs in cattle, horses, sheep etc. We know that all these animals will have some parasites, but we control parasitic levels with periodic medications to a level that does not cause clinical disease in our animals. The same is true for flea treatments we use on dogs and cats. We know that our animals will be exposed, so we treat and hopefully prevent these parasites from infesting our pets. In neither of these scenarios, have we considered breeding flea or intestinal parasite resistant animals.

Breeding for certain traits (color, size, speed, strength, temperature adaptation, etc.) are possible in animals over several generations. But parasitic relationships can be some of the most complicated processes in nature. Some parasite lifecycles involve multiple hosts' interactions and years to complete their lifecycle! Evolutionary time needed for resistance or tolerance to develop in a host-parasite relationship takes hundreds of thousands to millions of years.

Why is Varroa destructor so bad in *Apis mellifera*? Because in this bee species, **Varroa destructor is like fleas combined with Rabies.** A pervasive parasite with a nearly 100% mortality rate (without prevention and treatment). Incredible!

At the 2019 Apimondia in Montreal, I went to see a lecture by Tom Seely. Perhaps you have heard of him..? He is a beloved man that about every beekeeper wishes would be their grandpa. He writes and tells of wonderful stories and experiments involving honey bees in a Winnie the Poo like woods in upstate New York. He advocates for Darwinian beekeeping, with a plethora of insightful management techniques but one caveat involves killing any colony with high Varroa mite counts with no treatment for Varroa. The idea is that beekeepers would select colonies that are resistant to Varroa mites without treatments and therefore self-control mites in their area. What Seely also says at the beginning of his talk (I was there) is that this method would not be practical or effective for commercial beekeepers or urban/suburban beekeepers.

So, let us do the math. Using recent (2021) USDA estimates, there are about 2.7 million honey bee colonies in the US, of which 2.2 million are owned by commercial beekeepers (6). That leaves only 0.5 million, a small 18.5% of the total. How many of these colonies do not encounter other bees? Hmmm...Well that is hard-to-find exact data on, but we know all urban and suburban beekeepers would be out, and any "rural" beekeepers that live within a 2–3-mile radius of other beekeepers **or** any possible known or **unknown** feral hives should also be out. Who would be left in this ideal scenario?

As a biologist, I have respect for Darwin and his theories, but going "full Darwin" without the benefit of modern medicine and technologies, I would not be writing this article because I would be dead. I would have died of pneumonia in early childhood, and if not then, from a dozen other ailments since. Most of you would not be reading this for similar reasons. Life expectancy in the US in just 1900 was 48 years for women and 46 years for men (8). Again, if I were living back in the "good ole days" ... on average, my time would be up. Look at Africa for another example. Human life expectancies in many "developing" African countries have been in the 40's well into the 2000's but recent increases to life expectancies in the 60's are largely due to increased access to medical health services (9).

In 2022-23, Australia targeted, baited, and poisoned **feral** bees due to Varroa mite infestation in their failed efforts to eliminate the disease from the island continent (10). Whether you agree or disagree with this approach, their reasoning is because feral colonies can act as a reservoir to spread the disease with no practical way to test, control, treat or eliminate the parasite from these colonies.

Yes, we did this, we created a huge agricultural landscape on Earth to support eight billion and counting human souls. Domesticated, managed honey bees are not going anywhere. They are vital to the preservation of public health as honey bees are needed in the creation of much of the food we and our animals require to survive. This has been a great accomplishment but not without negative sequelae. Loss of habitat for wild animals and plants and globalization of trade leading to a globalization of diseases exposing naïve populations to originally isolated diseases are our main challenges. We work hard to prevent and remedy as many of these negative sequelae as possible.

The questions we should all ask ourselves are: Have we increased the quality and quantity of life for us and our animals and how can we improve our current stewardship? But going back 10,000 years to hunter gatherer days and actual wild honey bees is not a realistic possibility or a solution to any modern challenge.

Frame Five - Honey Bee Management

Beekeeping is an age-old art that has been around for thousands of years. The art of medicine has been around for quite some time as well. But only recently has a veterinary approach been considered in keeping one of our most important agricultural animals healthy. This frame looks at beekeeping, honey bee management, with the addition of veterinary insight.

How will more AI affect our honey bees...

The Veterinary Diagnostic Approach in Apiaries: The Exam

One recent semester I was inspired to voluntarily sit in with undergraduate sophomores (19 & 20-year-olds!) and take genetics. Retake genetics. As a trained biologist and veterinarian, of course, I have studied genetics in the past. Genetics permeates just about every aspect of medicine and biology, which I have utilized throughout my career. However, the last time I sat in a genetics' classroom was nearly 25 years ago... and it would be quite an understatement to simply say that much has changed in this field since the last millennium. Bees have been part of my inspiration to update my brain on the subject since genetics can play such an important role in honey bee health. With the volumes of information out there on honey bee genetics, I am attempting to equip myself to pick through the weeds of opinions and science on the subject.

It is important that we never stop learning. In these next few chapters, I would like to walk you through how a doctor, a veterinarian, works through a diagnostic process. My hope is to give you insider insight on how vets are trained to think in this scientific approach and how this process can benefit beekeepers and honey bees. This first chapter will focus on history, prevention, initial exams, and record keeping. Part two will focus on diagnostics and we will wrap up the section with treatment plans, medications, and return full circle to prevention.

History: Good doctors, nurses, and veterinarians are trained to get a good medical "history" about any patient before doing an exam or even seeing a patient. We will ask questions. This could seem like the third degree and some of the questions may even, at times, seem "dumb". But in our detective work, we will use open-ended questions as a technique to try to objectively learn about you, your operation, and your animals. This knowledge can help us best determine and fully understand what challenges a beekeeper and their bees may be facing. This also helps us establish the Veterinary Client Patient Relationship (VCPR) which is required by federal and state veterinary laws for us to legally practice medicine on an animal or a group of animals. These laws state that "the veterinarian has sufficient knowledge of the patient" and "is personally acquainted with the keeping and care of the patient".

So, what can a beekeeper expect to be asked? The following questions are examples of typical information collected in a honey bee operation medical "history". These questions can and should be asked before a veterinarian starts any examinations of colonies.

1. What type of beekeeper (backyard, sideliner, commercial) and beekeeping operation (single yard, multiple yards, how many colonies) are you?

2. What is the reason for the consultation? What is the desire/goals of the beekeeper for the visit? Are there any concerns, previous or current suspect pathogens?

3. What is the duration and severity of any concerns?

4. What are the current biosecurity practices practiced by the beekeeper? Are there any biosecurity plans in place?

5. What medications, feed, or chemicals are already in use at the bee yard/hive, including duration and dosage of any treatments applied?

6. How do you manage nutrition with your bees?

7. How do you manage Varroa with your bees?

8. Are any relevant hive records available?

9. Have you brought in any new stock lately?

10. Can telemedicine be used to facilitate our visit or follow-up?

Prevention

The old adage, "An ounce of prevention is worth a pound of cure.", is a medical truth that is widely demonstrated in honey bee health. With all animal species, veterinarians spend much of our time applying preventative medicine. From nutrition to vaccines to parasite control to management practices, we employ preventative strategies as a rule in house cats to cattle herds. With our bees, we certainly prefer to prevent or limit diseases before they occur. Knowing what preventative practices are already employed in an operation can help us determine which diagnoses are more or less likely. I know beekeepers like their privacy. I do, too. But understand that information exchanged between a beekeeper and a veterinarian in a legitimate VCPR is confident and privileged information and it is a veterinarians' role to give beekeepers the best advice to meet your bees' and your operations' needs. We will encourage practices that not only prevent or treat one disease but improve the overall health of your colonies in the short and long terms.

Exams

After a good history is taken, it is time to examine your colonies. Full or partial inspections will be necessary depending on the situation. It is nearly impossible to evaluate a situation without an examination of the problem in the context of its environment. This is where you can be of great help to the veterinarian. Ideally these exams should take place in-person in your yard with you and your veterinarian working together as a team. First-time visits and new conditions require on-site visits per VCPR laws. With the emergence of COVID-19, telemedicine has made some visits and follow-ups possible utilizing this technology.

Here are some points to remember:

1. While a veterinarian will likely have their own tools, it is best biosecurity practice to utilize your beekeeping tools (smoker, hive tools, even veils) in your yard. This helps to limit disease spread to or from your yard. Be sure to have tools ready to make the best use of time.

2. Veterinarians are great resources for making practical biosecurity recommendations that could improve the health of your colonies. Pick their brains on preventative medicine practices you could employ. Try to have areas where handwashing and boot cleaning can be performed.

3. Veterinarians may use disposable gloves. I know there is a wide array of opinions amongst beekeepers in using gloves, but veterinarians are well trained in how to use disposable gloves in ways that can limit disease transmission and may choose to employ disposable gloves in certain situations.

4. If a veterinarian is visiting your apiary for a sick hive or yard, they should examine that hive or yard last to prevent disease spread as much as possible.

5. Exams should be efficient but thorough. Quick, jerky or rough handling of frames should be avoided. Veterinarians will look to follow your method of working your bees, if you utilize a smooth, confident approach.

6. During an exam expect that veterinarians will want to run tests and take laboratory samples to confirm any tentative diagnosis made in the field. Confirmational objective data is always good.

7. Expect that veterinarians will keep records of the exam and any testing done. This is required by law. You may request copies of these records from your veterinarian.

8. Expect follow-up calls and/or exams from the veterinarian. This is part of the process and the law.

9. If AFB is suspected, the state apiarist should be called immediately.

Recordkeeping

Collecting medical histories and conducting examinations are about collecting as much relevant information as possible and using this information as tools to best improve our bees' health. One of the best preventative medicine tools, helpful in both history and exams, is maintaining good hive records. Maintaining a system of individual hive identification numbers or names is critical in developing accurate records, as well.

In my travels around the beekeeping world, I have noticed that this is an area we as beekeepers can improve on. There are plenty of recording keeping systems available in-print, on-line or even homemade for beekeepers to easily integrate into their bee management system.

The Veterinary Diagnostic Approach in Apiaries: Finding the Diagnosis

In my high school and early college days, I had the privilege of shadowing several James Herriot types of veterinarians. From flying over the countryside in some old jalopy to the next dairy farm (with no GPS) to rattling off mile-long prescription drug names, I can remember innocently (and somewhat naively) marveling at how they could possibly know and do "everything". I wondered if they received some special bolt of knowledge from the heavens upon graduation from vet school. After 24 years of being a veterinarian, the idea and expectation of knowing "everything" just makes me laugh. So how does one approach the diagnosis of all diseases in all animals? Well, veterinary school and life experience has taught me a few things...

Allow me to share some insight that you may find amusing. In veterinary school, I was introduced to the DAMNIT diagnostic system, a mnemonic acronym and perhaps a somewhat sarcastic commentary on the way docs may feel regarding the task of searching for diagnoses that may be ellusive, vague, deceptive, life- threatening, and even unknown. Each letter in DAMNIT stands for a certain category of disease and is meant to direct us through a systematic, scientific, diagnostic method for every animal and every disease situation we encounter. The categories are as follows:

Degenerative

Anomaly, Anatomic

Metabolic

Nutritional, Neoplastic

Infectious, Immune, Idiopathic, Infarct

Traumatic, Toxin

With the exceptions of Neoplastic and Infarct categories, I believe the other categories could apply to various honey bee pathologies. Note that "Idiopathic" is a major category. "Idiopathic" literally means the disease makes an idiot of us. CCD sound familiar? Unfortunately, even with our modern medical technologies some diseases in human and animal medicine remain unknown. Bee maladies are no exception, but we keep learning each day.

These DAMNIT categories allow doctors to first consider widely the possibilities and then narrow the focus on what may be causing the disease. We will develop what is called a "differential diagnosis" list or a list of diseases we consider to be the most likely cause or causes of the issue at hand. We will then decide which diagnostic tests will be helpful in ruling in or ruling out possibilities on our differential diagnoses list. Many diseases in humans and animals, including bees, have similar clinical signs and presentations. Performing diagnostic tests are objective ways to obtain data that will support the most accurate diagnosis or diagnoses, which will hopefully lead to the best treatment, recovery, economic savings, and/or course of action.

One hard rule I learned early on as a veterinarian, is that there is no rule that says there must be only one problem, one diagnosis, one pathogen causing the animal's disease. This fact, of course, further complicates the diagnostic process, but complete and accurate medical diagnosis is not easy. In honey bees it is certainly not uncommon to see multiple stressors and pathogens contributing to the collapse of a hive. For example, if a beekeeper says their hive died of starvation or "winter" …ok but why did they die of starvation? If bees simply died due to winter, we would have zero bees left after one year. There were likely other underlying cause/s.

An objective diagnostic approach can even be helpful during a post-mortem exam of a hive, or what we would call a "necropsy". A necropsy exam can be performed, and diagnostic samples may still be taken to determine the cause or causes of the hive's demise. Finding a definitive diagnosis will be helpful in taking steps to preserving any remaining hives and may guide the beekeeper on necessary management changes.

Of course, experience working hives measured in both years and "mileage" can be invaluable in making the most accurate hive assessments. Combining beekeeping experience with the best objective data gathering is a win-win for beekeepers, bees, entomologists, and veterinarians. Much data we currently have about hive death is based on survey data, often utilizing non-specific diagnosis categories, and educated guesses from a variety of levels and types of beekeepers. Continuing to "team up" on ways to obtain the most accurate data will improve our collective ability to assess what is objectively happening to our hives.

Get a handle on the normal look of your bees

Diagnostic tests

Diagnostic tests can be performed in the field or in a laboratory setting. Field tests are performed first and may be good for screening for disease and/or developing tentative diagnosis/ses, but they rarely absolutely confirm a diagnosis. Confirmatory and gold standard tests are typically sent out to a laboratory.

Field tests

Field tests start in the bee yard with exams or hive inspections. The exam itself is a diagnostic test and a very valuable one. Exams in honey bees start with a visual inspection of the exterior of the hive. Much can be learned without even opening the hive, including, the strength/population of the hive, hive activity/behavior, normal morphology of adults, the presence of dysentery and/or any abnormal dead bees at the entrance. If scales are utilized, hive weight and hive weight patterns over time can be assessed.

Very few diseases have what is called a "pathognomonic" sign. A pathognomonic sign is a clinical sign found on a physical exam that is absolutely diagnostic for a specific disease. A couple of bee diseases do have pathognomonic signs, one is "Chalkbrood", caused by the fungus *Ascophaera apis*, is seen at the entrance of hives as discarded, white, mummified brood.

After a thorough exterior exam, an internal exam should be performed. Ideally, internal exams should be avoided in harsh or cold weather conditions. Four major items should be assessed while in the hive: nutritional status, queen status, brood status, and adult bee status.

The hive should be assessed for adequate stores, both honey and pollen. Brood frames should be evaluated for even and seasonally appropriate patterns to assess both the health of capped brood, open brood, and the queen. Adult bees should be observed for normal behavior and anatomy as well. As much of the hive as possible should be examined to gain an adequate and thorough evaluation of the hive's health status. Many brood diseases and viral diseases, affecting both brood and adults, can have similar clinical signs. Varroa mites should always be considered as a primary, underlying, or contributing cause of hive abnormalities.

If brood disease is suspected based on hive assessment further diagnostics should be performed. The Match Stick test and Holst Milk tests are field tests that can be performed quickly with few materials to help determine if AFB is to be suspected. Commercially available antibody tests for EFB and AFB are also available and can be performed in the field.

Varroa mite count should be performed on a routine basis (monthly to every other month during the beekeeping season) during regular inspections and with sick hives. Alcohol washes are the preferred method for accuracy. Nosema spores can be detected by collecting whole adult bee samples for later examination in the lab. Whole honey bee samples are macerated and then viewed under a microscope to detect Nosema spores.

Confirmational tests

Confirmational tests are sent out to laboratories. The USDA Bee lab in Beltsville, MD provides testing for foulbrood as well as antibiotic resistance screening on samples. Viral testing can be found at a few university labs in the US and Canada.

Treatment- what if it cannot wait?

Ideally, it is best to have a proven diagnosis before treating any disease. Using drugs/chemicals haphazardly can lead to further harm to the bees and can lead to drug resistance. However, there are a few exceptions, namely, American Foulbrood. If AFB is suspected in the field immediate action and treatment should ensue. The state apiarist should be contacted, and state and federal laws should be followed for treatment. Veterinarians are permitted to start treatments immediately based on a tentative diagnosis if they find it appropriate. State apiarists, the beekeepers and veterinarians can all work together to be sure that proper treatment is initiated, and samples are also sent off to a lab for confirmation of AFB.

The Veterinary Diagnostic Approach in Apiaries: Treatment

In our big blue world, there are nearly 1500 different infectious diseases known to affect humans. 60% of these diseases are **zoonotic**, that is diseases that are transferred between animals and humans, and 75% of new and emerging infectious diseases are zoonotic (1). Officially, there are no zoonotic diseases in bees (with the rare exception noted in the literature of a few poor souls who decided to inject honey into themselves, which happened to be contaminated with *Paenebacillis larvae* or American Foulbrood, AFB, spores) (2). Every other domestic animal and many wild animals can transmit disease to us, so the lack of zoonotic disease in bees is quite unique. This lack of natural, direct, zoonotic disease transmission between bees and humans allow beekeepers to enjoy a more relaxed, low-risk interaction with our bees regarding disease.

However, our bee's health and the drugs we use to treat them affects us all - humans, other animals, and our environment. The concept of human, animal and environmental health being inevitably intertwined is called "**One Health**". Our bees' good health is vital to provide pollination services for a large portion of our food supply. The drugs and chemicals we utilize to treat bee diseases can leave residues and contaminate bee products, like honey and wax. **Antibiotic resistance** is considered by national and global health organizations to be a top priority, public health crisis in the world.

700,000 human deaths are attributed to antibiotic resistance *every year* (3, 4) Overuse or misuse of medications can lead to treatment resistance to the particular disease within our patient, our colonies. One should always consider the overall effects of developing treatment plans and administering drugs, medications, pesticides, home remedies- all chemicals, to your bees. Below is some guidance on how veterinarians can help beekeepers employ treatment plans.

The Development of Treatment Plans

Ideally, to have the best outcome for any patient and to avoid drug resistance, it is best to have a proven diagnosis before treating any disease. As a young veterinary student, I can remember being eager to learn about how to treat diseases - how to fix it! I would sit in lecture halls for weeks on end learning about various pathological disorders…how they look, how they develop, even how they smell, but every time, near the last lecture of the term, when treatment would finally come up for discussion, the same old phrase would appear, "Treatment: Depends on the Diagnosis". Everyone wants a magic pill to fix all our issues, but it is hard to fix what you have not identified. Veterinarians *are* permitted to start treatments immediately (pending confirmatory testing) based on a tentative diagnosis, if they find it appropriate, such as in a case of suspected AFB.

Define Drug Resistance

Over the years, I have found that there is some confusion about the nature of **drug resistance**. So, let us take a moment to step back and define drug resistance. Antibiotic resistance is a specific example of drug resistance, but resistance can occur with any chemical that is used to treat an infectious agent or pest. For example, there is concern about resistance development with miticides used in the treatment of varroa mites. Drug resistance means that a given population of bacteria, varroa mites, or whatever, have been exposed to a certain drug. The drug kills most of the pathogens, but a few may survive. These lucky few have a genetic predisposition that allows them to survive and go on to create the next generation of bugs. This new generation of drug resistant organisms emerge, and our drugs become less useful. Drug resistance is **not** the patient "getting used to the drug". Now that's cleared up…A careful review of any medications previously used in your bee yard/s could inform best practice choices for your bees going forward.

Employment of IPM Practices

Integrated pest management is a practice commonly employed by beekeepers and veterinarians alike. Most diseases of animals, including bees, are not treated with drugs alone or proper treatment can be achieved without drug intervention at all. Engaging in best management practices, re-queening, proper nutrition, and sometimes just plain rest are often the best medicine.

Drugs are only one tool in our toolbox for fighting disease. In fact, of the few bee drugs that require veterinary intervention, their effectiveness is only in part to nearly useless. Since antibiotics are not effective against AFB spores, burning is required or highly recommended for hives infected with AFB, and for *Melissococcus plutonius*, European Foulbrood (EFB), treatment with antibiotics is typically only used in severe cases and in conjunction with other IPM practices.

Good News: Most backyard beekeepers employing good biosecurity protocols should not have to use antibiotics because typically their bees are kept in one area (reducing stress and exposures). Also, luckily, AFB is a great example of an **endemic** disease. What is an endemic disease? Endemic disease is a disease that is ever present in a geographical area but typically in low, manageable levels. Due to the long surviving spores that AFB produces, soil is contaminated. While data of AFB incidence can be difficult to obtain due to the stigma of this deadly disease, known prevalence data year to year only affects a small fraction of colonies. The literature supports that it is likely that many healthy hives exposed to AFB spores can manage the infection sub-clinically and remain asymptomatic.

Provision of VFDs and Prescriptions for Antibiotics

Since 2017, the FDA has required a veterinary feed directive (VFD) or prescription from a veterinarian to administer medically important antibiotics to bees. The formulary is simple and limited. There are three approved drugs for use in bees, Oxytetracycline, Tylosin, and Lincomycin, available in 11 different approved veterinary preparations. Only one drug, oxytetracycline, is approved for the treatment of *Melissococcus plutonius* (EFB) in either a VFD or prescription. For AFB, only oxytetracycline is approved in both VFD and prescription forms. Tylosin and Lincomycin can be used for AFB only by prescription.

Not every vet will agree to see bees. To better serve their clients, many veterinarians self-limit themselves to the species of animals they serve, because the scope of what we may cover is so vast. This is not uncommon and not intended to be exclusive. You would not call a cat clinic to see your goat, right? But how do you find a vet for bees? Here are a few suggestions. If you already have a local vet for your dog or horse or whatever, it does not hurt to ask if they are willing and able to see bees. The Honey Bee Veterinary Consortium (HBVC) has lists of vets by state that are willing to see bees. Your state's Department of Agriculture (DA) has state veterinarians who may have contact lists of vets willing to see bees. For example, in PA, I consult regularly with the PADA on bees and compiled a list of vets interested in seeing bees.

Proper Use and Withdraw Times

I cannot stress enough the importance of proper use of medication. Labels and indications for use must be followed. It is the law, but also good practice in providing the safest course of treatment for your bees, preventing drug resistance, and preventing residues in bee products. Many drugs, like antibiotics and miticides, cannot be used when honey supers are on or must be withdrawn 4-6 weeks before honey supers are added. This requires careful management planning and record keeping to achieve. For antibiotics, approved indications of these drugs can be used for prevention, control, and/or treatment of disease. This is of primary importance in the management of foulbrood in migratory and/or commercial operations.

Holistic Health Practices: Beyond Antibiotics and Foulbrood

While AFB and EFB certainly can be serious disease issues, their prevalence and overall impact pales in comparison to nearly every hive in the US and much of the world being threatened by the panzootic (pandemic), varroosis. Good health requires evaluation of the whole patient. Would you only see your doctor for two diseases and treatment with three drugs? Thankfully, most backyard beekeepers may never have to deal with foulbrood, but many backyard beekeepers could use more help understanding how to manage the overall health of their bees and maintain strong colonies.

Certainly, state apiarists, university entomologists, seasoned beekeepers, and various beekeeping organizations offer many great resources for beekeeping. These groups and general beekeepers alike are still looking for more support to cover the demand for good information and services. Here's where bee savvy vets can add their abilities to partner with the industry.

I think the stage is set-up for some complimentary relationship building. Veterinarians that limit themselves to small, companion animals would be well acquainted with the nature of the relationship backyard beekeepers have with their bees. Large animal veterinarians would be well up on the curve with working with commercial beekeeping farmers. Within the profession there are also research and lab animal vets that can fit in quite well with entomologists.

Continuing education on bee health is included in nearly every major veterinary conference since 2017 and veterinary schools across the US and Canada are adding bee curriculum. Currently, I am working with a group developing a veterinary textbook series on bee health and I am writing the chapter review on registered drugs used in the US and Canada for honey bees. This review will include not just antibiotics but other medications for use in bees for a variety of medical conditions.

The Best Treatment: Prevention

As I end this three-chapter series, I hope it gave you some insight into how veterinarians are trained to approach disease and help achieve health in our patients. And with the end, we are back to the beginning...the best treatment is prevention. Prioritizing the four key elements of good husbandry/understanding bee biology, varroa and other disease control, nutrition, queen status/genetics will go a long way in keeping the healthiest honey bees.

If We Only Followed the Directions!

The Difference Between a Poison and a Remedy

Most people hate reading "the directions". Who wants to waste all that time when we can jump in and just figure it out, right? Following the directions can be even worse. Just today, I was floating through my lab checking on students who were working on an assignment. When I asked one group how they were doing, they said "Fine, but we started on Exercise 1 and then realized the handout said to start on Exercise 3. So, we've wasted 20 minutes...if we only followed the directions!" Like teaching, medical directions are no exception to poor compliance. Most people do not like taking medications or giving medications to their animals. In fact, studies have shown that medical compliance with doctor's orders is systemically low with up to 75% non-compliance (1). This trend is mirrored in veterinary medicine as well (2). I get it. I would prefer not to prescribe chemicals for my patients if I could avoid it. Unfortunately, many of the disease challenges our bees face do not allow us that utopian luxury.

Let me be clear about the decision of administering any drug, medication, or natural remedy treatments. No matter how they are labeled ("hard", "soft"), they are all chemicals. Antibiotics are chemicals. "Natural" products are still chemicals. Even "soft, natural" products come with a lengthy package insert of their chemical hazards, can kill bees/queens, and harm us, especially if they are not used properly. I do not care to take sides or discriminate based on common treatment categories. IPM is always part of the treatment plan considerations. My goal is to use whatever works with the least number of side-effects based on clinical evidence observed and documented in the bee yard. But if I could encourage beekeepers to follow one direction when administering treatment/s to your honey bees it would be, "Please, use the correct *dosage*".

It is All about Balance: Considerations with Dosage

"All things in good measure" or "moderation" is good advice. In medicine, dosage is defined as the amount of a substance given to a certain patient over a period of time. The purpose of utilizing the correct dosage of a drug/chemical is to maximize treatment efficiency while minimizing possible side effects. Dosage can be further tailored according to the needs of a specific patient in a specific circumstance. Realize no two hives or bee yards are the same and may require different interventions. Since Varroa mites are the single, biggest health threat to our bees, I will often use Varroa treatments as examples throughout this chapter, but the principles outlined here can be applied to any medical treatment for any pathologies. The following are important, practical considerations for beekeepers to master in determining dosage:

Timing: Most honey bee treatments are designed to be given during certain seasons and certain weather conditions. Understanding the biological rationale for when and why we use a treatment is critical for success. For example, oxalic acid treatments for Varroa mites do not penetrate wax cappings (where most mites reside), can kill open brood, and are thus intended to be used in broodless times. I am a big fan of oxalic when it is used properly. But if a beekeeper is employing oxalic acid multiple times during brood rearing times, (which is most of the main beekeeping season from about February through October in my PA neck of the woods), this treatment is unnecessarily exposing your colony and queen to a chemical that can acutely or subacutely effect bees and brood, and has low effectiveness on the mites at the time. Please be sure you have a plan as to when and why you may use certain treatments during the beekeeping season.

Amount/volume/formulation: I must admit I am a "to taste" cook. I sort of follow recipes, but I like to add in my own amount of spice or variations. Does this sauce need more wine? Of course, it does! ... at least two more "pours" ...fun stuff. But in medicine, measuring the actual amount is critical to treatment success. This is the *dose* of medication. It is also critical to understand the concentration of the medication of different formulations of the same chemical may have different compositions and therefore amounts given. More is not better, that approach may be toxic. Less is not better, that approach may cause the medication to be less effective and promote drug resistance. In some cases, after careful evaluation and especially in weaker patients (colonies), veterinarians may adjust doses to the specific need of the hive.

Number of Doses/Frequency: I can remember when I used to make fun of those "old people" pill boxes with M-F designations. Now I have them in my house. Missing doses is a common reason for treatment failure. Remember a treatment is not complete until all the directed doses are given with the correct amount of time in-between. There are good reasons for this. The number of doses and frequency of medication administration is often based on the lifecycle of the pathogen and/or the half-life of the chemical (essentially how long its effect lasts) in the patient. Be sure you mark the calendar for your bees' meds, too.

Duration: The duration is how long a treatment plan lasts. Some regiments for bee treatments can be long and admittedly annoying...7 days, 21 days, 42 days... but like the number and frequency of doses, the duration is often based on the lifecycle of the pathogen and/or the half-life of the chemical in the patient. The important thing is to be aware of the necessary duration and do the "math" before you apply the treatment to your hives. What your plans are for your hives in a week, a month, two months in the future may determine the best treatment choice.

Application method and distribution: In apiculture, there may be different choices for delivery of a medication to our bees: patties, dribble, vapor, strips, in sugar syrup, sugar dusting, etc. With application four things should be considered:

1. Is this the best treatment delivery to get adequate distribution of the medication to the bees?

2. Is this distribution method the safest/least toxic to bees?

3. Am I comfortable using this application method properly?

4. Is this method the most cost effective for my operation? Do some research here to be sure you are on the right track.

Expiration: Using medications that are expired may accomplish three things. Expired drugs are less effective in treatment, may contribute to resistance, and/or they break down and their metabolites can become toxic. Interestingly, many drugs after expiring usually just become less effective and not toxic. However, tetracyclines is one category of antibiotics that are known to increase in toxicity after expiration. Coming from a farm mentality, I get frugality. But believing that using expired drugs is working to treat or prevent disease and/or is economical is a dangerous fantasy.

Withdrawal times: Most drugs we use in animals have a time when food products cannot be used from that animal after a medical treatment. In honey bees, many medical treatments we use can have post-treatment time and some even a pre-supering timeframe, in which we cannot have honey or supers on our hives for honey intended for human consumption. Some of these times can be up to 6 weeks. Again, this takes careful pre-planning of your beekeeping season to be sure you are not contaminating your bee's honey.

Combinations of meds: If you are a horse person or manage any kind of herd animal, you are likely familiar with "strategic deworming". This means we know our pasture animals are exposed to parasites in the environment, so therefore we employ multiple drugs in "rotations" over the season to keep the parasites at a manageable level and avoid drug resistance development. We can monitor the effectiveness of our drug rotations utilizing periodic fecal egg counts. This is becoming our pattern with Varroa management. Please do not think that simply treating once for mites with formic acid or amitraz or thymol or whatever, and then you can check the box for varroa for the season. Most beekeepers are now employing multiple treatments over the beekeeping season at appropriate times to keep Varroa in check. For example, amitraz in the early Spring or Summer dearth, formic in the Spring/Summer (if it is not too hot), oxalic in the late Fall/early Winter. The only way to determine if your treatment plan is working is to do regular mite counts or other monitoring diagnostic for whatever disease or pest you are managing for. Our goal is still to use the least amount of chemicals for the highest effect. However, that formula may be a treatment plan involving several drugs. Using "more" effective drugs appropriately could mean using less overall.

Evidence of treatment success: Doing regular quantitative mite counts is the best thing you can do for the health of your hive. Learn how to do an alcohol wash correctly. I know, I do not like killing bees either, but 300 bees are a diagnostic sample that could save the life of a hive and tens of thousands of bees. Many experts recommend doing counts monthly during the active beekeeping season. Personally, I think three counts a year is the minimum (Spring/Summer/Fall). If you have not done mite counts before, start with three and work your way up to what is manageable for your operation. Pre-treatment and post-treatment counts are best for evaluating treatment effectiveness.

Records: All medications with all the above considerations used in your bee yard must be recorded. This is the biggest favor you can do for your bees, yourself, and any other mentor beekeepers or veterinarian coming into your yard to assist you. Develop a system you can understand a year from now.

Purpose: In all these things, beekeepers must consider ultimately what they want to achieve with their bees. Depending on if you are a backyard beekeeper, a commercial beekeeper, a honey maker or migratory beekeeper, your goals, timeline, and environment will all differ and will impact the most effective treatment regimen for your bees.

Consequences of incorrect dosage

Ok, so that is a lot to consider for following "just" the one direction of correct dosage. However, avoiding serious consequences can be reduced or avoided by keeping the above in mind. In summary, these consequences could include:

1. Treatment failure. All that time, money, and chemical exposure for nothing.

2. Queen effects. Death of the queen or sublethal effects that reduces her performance. Remember the queen lives longer than all other casts, so she must endure more treatment exposures.

3. Drug resistance development. Unfortunately, these poor choices end up affecting us all.

4. Death of the hive due to succumbing to the disease (with underdosing) or treatment toxicity (with overdosing).

Keep in mind that all these treatment failures can have significant economic impact on the beekeeper.

One last point I will leave you with to ponder, pesticides. While I am not about to jump into the ring of discussion about how much impact pesticides have on our bees, I will point out that the highest potential concentrations of pesticides and other chemicals in our hives are often the ones we use. Equip yourself with knowledge, pre-plan, and choose wisely when using any chemical in your yard.

"Fall Planning"

Of all the seasons in beekeeping, I believe Autumn may be the most underrated in its importance in honey bee health. Much to do is made about "Winter losses" and more currently, "Summer losses", but what about Fall? We hear less about what goes on in a hive during the changing colors of Autumn. Kudos to the beekeepers who have coined the phrase, "Take your losses in the Fall." Seasoned beekeepers, farmers, and other agricultural animal caregivers understand the importance of always thinking several seasons ahead of where they are...to ask where are we going? What are the goals? *Now* is a fleeting moment, but many of the decisions we make now can have lasting effects in the next season/s.

Why is Fall so important? Because that is when the major health indicators and challenges to a hive (varroa, nutrition, queen status) are peaking while the colony is entering the Winter season. For many geographical areas in the US, Winter is the longest and most harsh environmental stressor to our bees. Fall is also the last opportunity you get to have a meaningful and thorough good keeper interaction with your hive/s for months, (as we tend to not open our hive up for much inspection in the winter). If you can execute a successful Fall health plan for your bees, you will not only provide them a chance to get through the Winter, but you will also set up your bee yards for success the following Spring. The following outlines key components to include in your honey bee fall health plan. A good time to incorporate the components of the plan is at Fall extraction (if that applies to your situation).

Fall may be the most important beekeeping season

Honey Bee Fall Health Plan

1. Varroa treatment and testing.

Take a mite count in October. You must. This is something you need to see. Please do not just take a count in May and think you have checked the box for the year. Fall counts will likely be high. Like 30 in an alcohol wash. Do not panic. This is what varroa mites do. They reproduce... *exponentially*...and their population peaks quickly in the Fall. Now you have knowledge, and you must treat (even if you treated before).

After any honey supers are removed, oxalic acid is a good go-to for Fall. Dribble or vaporization may be used. Vaporization allows for treatment without opening the hive. Most beekeepers will utilize three sometimes four, once-a-week treatments in November and even early December (pending the weather) to take advantage of the natural seasonal brood break and cover any late brood emergences. Remember the queen slows and stops laying in late Fall through late Winter. Also, any remaining worker brood (with a 21-day development cycle) will be "hatched" within the three-week treatment period. Oxalic acid is not effective under wax cappings. Be consistent with your treatment in every hive in your operation and your colonies will be off to a good start for the next Spring.

Check the hive's nutritional status

2. Know the Queen status. Take your losses now- combine.

During Fall extraction confirm the hive's queen status and the general health of the hive. If the hive is weak and/or not queen right, it is likely best to cut your losses now. Depending on the situation, you may be able to harvest the honey and combine workers with other hives. Finding a replacement queen at this time of year will be difficult and probably costly, and there is no guarantee a new queen will be able to "fix" whatever the problem is, especially on a short time table. Obviously if you only have one hive, this can be a problem. Making those Spring splits could now come in handy. Again, planning often starts seasons ahead.

Remember honey bee populations within a hive normally fluctuate over the year, peaking in the Summer and diminishing in the Winter. Hives can follow the same pattern. Strong colonies have a much better chance to making it to Spring and by April you will be able to make multiple splits from one overwintered hive.

3. Provide proper nutrition – leave some honey.

Depending on where you live in the US, honey bee colonies need 40-80lbs of honey to have enough energy and nutrition to get through the Winter. Be sure that you are giving it to them. Honey is their perfect food. This means leaving one or two supers on a hive for winter. I understand many honey operations cannot afford to do this, but if losses are primarily attributed to "starvation", perhaps some economical math should be done to evaluate the situation further. Are the hive losses, supplemental feed, and additional labor worth it? I cannot imagine a cattle farmer providing half the hay needed for winter to their herd and then coming to terms with a 40% loss over the winter.

Certainly, you can check on your bees in late Winter/early Spring, as possible, to see where they are and perhaps provide sugar or pollen supplements but setting them up well in the Fall will go a long way.

4. Good shelter.

Fall is a great time to inspect hive boxes to be sure they will make it through the winter. They should be sturdy and free of cracks or leaks that could let rain or snow melt into the hive. Maybe they just need a quick paint job. If you find any defective boxes, Fall is a last opportunity to switch them out so you can have cozy bees before the weather turns cold. Consider adding insulation to the top covers at this time. Constructing a wind block to be in place before winter winds set in should also be in your on-going plans.

5. Last/less inspections.

Your last hive inspection should be done in the Fall, again at Fall extraction is a good time. If your colony is queen -right and looking strong, say a prayer, get out, and stay out until Spring. Now is not the time to smash the queen. If she is present in October and you take care of the previously mentioned items above, she will almost always reign through the Winter. Your biggest problem will be catching her swarm in the Spring.

6. Other pest/disease control.

While not practical for large beekeepers, I would recommend that smaller scale backyard beekeepers invest in a chest freezer (or plug the one in that has been sitting in the basement), to store extra frames over the winter. Freezing frames kills and prevents a wide variety of pathogens including hive beetles and wax moths. If you do not use mouse guards all year long, Fall is the best time to install them. So, this Fall while enjoying some Autumn honey, incorporate a Fall health plan for your colonies, and set your bees up for success next Spring.

A Surgical Approach

Did you ever have a moment within the crazy pace of life where you suddenly stopped, looked around, and asked yourself how (or why) did I get here? I find myself asking this question now that I seem immersed in swarms of all stuff bees, something I would never have imagined 10 years ago. What is the appeal? I suppose for many folks the reason or reasons for becoming a beekeeper are many and different: I wanted a hobby, my grandpa did it, I like to be outside in nature, I want to "save the bees," I like honey, I want to make money (ha-ha), etc. But it recently dawned on me that it is surgery. I am a surgeon. I do not mind getting my hands dirty. I like to examine things, take things apart, find the problem, hopefully fix the problem, and put things back together for a positive outcome. That is beekeeping.

Really, think about it. A honey bee colony is a superorganism. Every time you sever the propolis barrier and open a colony you are looking at its guts. Donned in specialized PPE analogous to a surgeon, beekeepers review the workings of distinct parts of the organism's systems. During an inspection procedure, various parts of the hive normally unseen are revealed and more closely examined. Queen status, patterns in the brood and energy stores are evaluated. This assessment is a review of reproductive systems, metabolic systems, excretory systems, immune systems, and even communication systems analogous to the endocrine and nervous systems within any mammal. Any abnormalities are noted, removed, cut out, or treated with medications. Parts of the hive can be manipulated to put the colony into a healthier position. "Donated" parts from other hives can be added to colonies to replace damaged or missing components. Samples of tissue may be taken for definite diagnosis. But don't take too long, a patient's ability to thermoregulate can be affected by prolonged procedures. If you are a careful surgeon, you will only damage a few cells (smash a few bees) on your way in and out and the colony will heal any cracks in a few days. Many times, surgery can not only diagnose a problem, but it can also be the cure. Other times it can just provide the most accurate prognosis, whether good or grave. So, am I talking about beekeeping or surgery? Hmmm...like I said.

If you have had some struggles with keeping your bees, consider looking at it from a surgeon's perspective. No matter the subject, the details and intricacies of any topic can often seem complicated, but in both surgery and beekeeping, understanding and mastering a few, consistent practices can go a long way.

I have been hesitant to share this because I am sure as soon as I say it, bears will come and destroy all my honey bee colonies, and I do not want it to come off as a brag, even a humble one. However, so many beekeepers suffer 40% losses every winter and I have been asked to share and summarize what I do by multiple beekeepers... so here goes...I have had 100% winter survival for 3 winters running. I started with one colony in 2019 and now I manage eighteen colonies as of yesterday (it is May 2022). I could have many more colonies if I were actively trying to make more splits.

Disclaimer: This may not work for everyone in every situation, much may change or need adaptation regarding geography, operation size, and purpose, but I am happy to share what has worked **for me** to this point and that I am starting to give away bees because I have too many. If this information helps you, awesome, if not, please turn the page to the next chapter.

Know thyself

With degrees in both biology and veterinary medicine, and decades of agricultural animal care experience... I **still** studied honey bees for 2-3 years, observed dozens of other beekeepers and scientists, the good, the bad, and the ugly, **before** I became a keeper of honey bees. This certainly does not mean you have to have multiple degrees or certifications to be a good beekeeper, but it sure does help to spend considerable time researching what you are doing before you start doing it. You must understand anatomy before you can do any cutting.

Biosecurity

I only perpetuate or raise my own stock and queens. Period. I have never brought any outside bees into my yards. Only new equipment is used, and equipment is not shared outside of my yards. There is double electric fencing around my yards. Mouse guards are in place all year round. The local police are my friends. I wash my jacket once in a while.

Nutrition

I leave some honey on... one or two honey supers over winter or at least 60 lbs. (I live in Western Pennsylvania). It is their best food. The hives are located near water that is available year-round. Dry sugar is fed in January and pollen patties added in February, whether they need it or not. All supplemental feed is removed in April. A 1:1 sugar syrup may be fed during dearth, if needed or just leave some Spring honey on.

Sharing food inside the hive

Medical management

Varroa. I test hives with an alcohol wash at least 3 times a year; Spring/Summer/Fall. This allows for treatment adjustments, if needed. Currently, all my hives are on this Varroa treatment plan: Apivar (amitraz) in summer, after Spring extraction and before Fall supering, and OA vaporization, three times once a week in November and maybe a fourth time in December, pending weather. More recently, I have sometimes added an early Spring treatment, like Hopguard, if counts are too high in the Spring. I remove drone brood, if it is convenient, during a hive inspection. Between swarms and splits there are natural brood breaks in my hives.

Antibiotics are never used on my bees.

I keep hives on gravel or another dirt barrier. This can reduce several different pests.

Hives must be checked at least on a weekly basis all year round. This means mostly external exams in the winter but during the beekeeping season, internal inspections may be required weekly. You must know the when, how, why in doing a proper hive inspection and what it means. If you do not, please do not keep honey bees until you do.

General management techniques

These are the things that beekeepers will talk about until they eat all the donuts and drink all the coffee. What is "best" probably depends on your given situation. Here are somethings I do:

1) Top entrance

2) Solid bottom board

3) Insulation board inside the outer cover of hives in the winter; consider wind blocks that may work in your area

4) Limit pesticide use in yards

5) Freeze unused drawn comb frames

6) Keep detailed records on hives

7) Overwinter hives in 2-4 boxes depending on colony size, deeps, or mediums, 8 or 10 frame...whatever

Other disclaimers

I am not doing this for a living. I have a small sample size. I have not done this for 50 years. (Although my great grandfather had a few hives at one point. Does that make me a 4th generation beekeeper?). I am not perfect. I have lost swarms and a couple of queens/colonies during the summer due to robbing or failure to re-queen. I hope this sharing of best practices can help you in planning out your operation!

Honey frame

Frame Six - Honey Bee Medicine

Honey bees deserve to have a doctor. In the last several years, advances have been made to elevate our beloved honey bee into the realms of what we consider standard of care for our other animals. We are developing new treatments for the improvement of their lives and the betterment of their deaths.

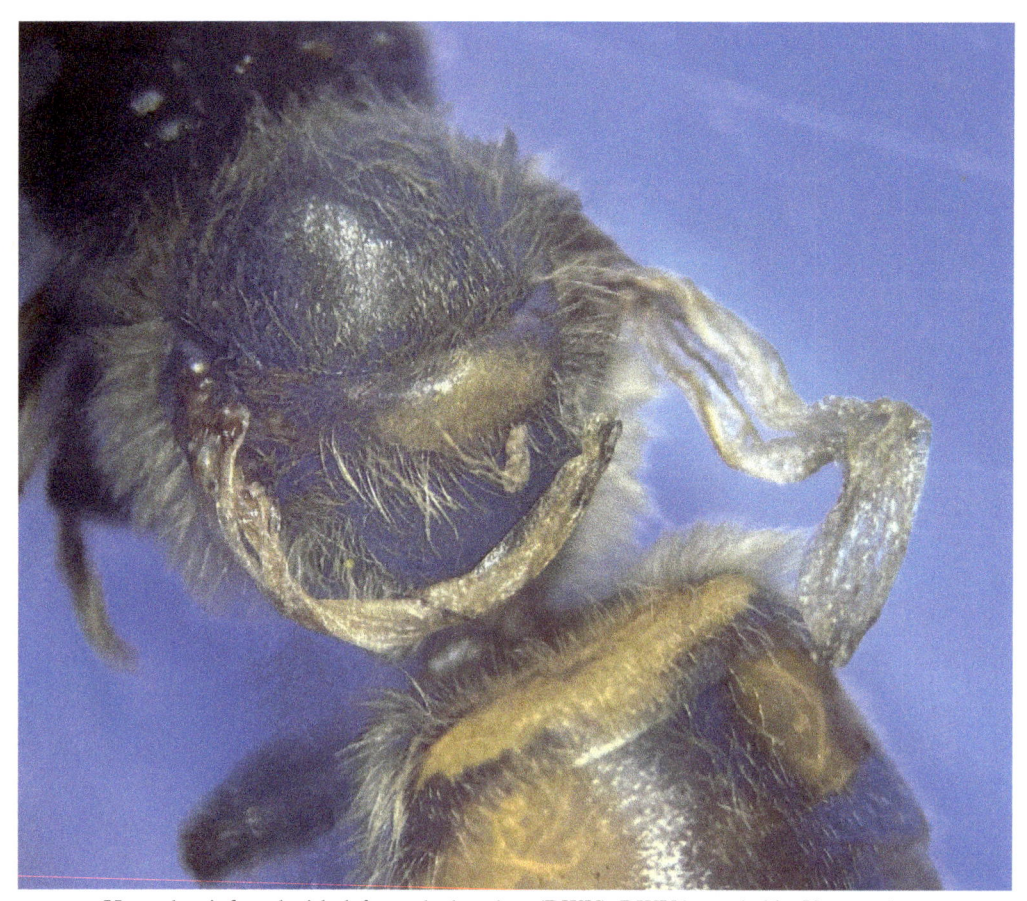

Honey bee infected with deformed wing virus (DWV). DWV is carried by Varroa mites.

A normal cleansing flight on a warm winter day

Diagnosis Is a Tricky Thing

During my early years in clinical practice, I managed a case of a dog, "Red", with recurrent vomiting. Red was a fiery brown, mixed breed pup owned by a happy young couple. The dog seemed otherwise healthy, but he would come in to the hospital for vomiting, we would treat him with the typical meds, he would improve, but ultimately, he would be back in a few days. X-rays and bloodwork did not show anything particularly abnormal. But one evening, Red came in very sick, and I decided it was time to open him up. X-rays were now consistent with an obstruction of the stomach. Interestingly, in surgery, I found a wine cork in Red's stomach at the pylorus, blocking the outflow to his duodenum (small intestine). I was able to easily "pop" the cork from the stomach, much to my nurse's amusement. After removing the cork from the stomach, I continued to "run the gut" ... by slowly palpating the entire small intestine.... duodenum, jejunum, ileum, down through the colon, feeling for any other foreign object, lump, or bump. My nurse asked me why I was doing this since I clearly already found the "problem". My answer was because I was trained to always look for more than one (condition).

Diagnosis is a tricky thing. Becoming a good diagnostician requires years of training and experience, knowledge, the curious inquiry of detective work, and the ability to see the big picture. Coming up with the "correct" diagnosis is a challenge to all doctors, including veterinarians. With colony loss being such a vexing issue in the beekeeping industry, beekeepers certainly share correct diagnostic challenges to identify and understand what happened to lost hives.

Medical conditions and diseases are diagnosed by meeting certain set criteria for diagnosis. For most conditions or diseases, this includes finding **clinical signs consistent with the disease or condition *and* a positive diagnostic test.** We rarely diagnose disease in asymptomatic patients, as it is hard to find something that does not present itself.

In beekeeping, we often use state and national beekeeping surveys to study reasons for colony loss. I would encourage beekeepers to take the time to fill out surveys. Better data in equals better data out, which will benefit all beekeepers.

However, all scientific surveys, while helpful, have innate limitations. With beekeeping surveys, returns are often 10% or less of all beekeepers, and reasons for losses are often educated guesses from a variety of types and levels of beekeepers. Diagnostic testing may or may not be employed to confirm disease diagnosis. Sometimes survey questions can be hard to answer because they do not apply to your situation. I recently filled out a survey that did not give an option for no loss over winter. I was forced to give a reason for loss even with none. Reasons beekeepers select for colony loss may be somewhat ambiguous with options like starvation, winter, summer, queen issues, weather, and various environmental hazards (chemicals, pesticides), other, and I do not know. For "starvation", ok but why really? Were they weak in the Fall, full of unchecked mites, did you leave them 40-80lbs of honey? For "winter", ok, but if all bees died because of winter, we would have zero bees in one year....so what is the real reason/s? Look for more than one.

Sometimes even "diagnostic" words do not help much. In medicine, we have a sophisticated sounding term for "I don't know", *idiopathic*, literally meaning the disease makes an idiot of us. In beekeeping, it is used to describe idiopathic brood disease. We still have much to learn.

Last summer I lost a hive. It was a split that I made from a very robust hive at Spring extraction. The hive made a new queen, she made it back from mating and was starting to lay, when the summer dearth and the worse drought in decades hit. I tried to feed in-hive sugar water to all hives, reduce entrances, but to no avail. The stronger hives took out this new queen and her colony by relentless robbing. So, what killed this hive? Robbing? The dearth? The drought? A beekeeper who gambled with a weaker split in a stronger yard and lost? Many would say "robbing". But what caused the robbing? All these things coming together contributed.

The problem with diagnosis is it is usually **not just one underlying cause**. In most cases, it is multiple things that bring down a hive or any animal. Co-morbidities. We all now know this word "co-morbidities" due to COVID-19 but how underlying conditions weakens any animal or human to disease or death has been well understood for a long time. Every stressor can contribute to colony loss. One stressor may not take out a healthy hive but persistent, multiple stressors over time certainly will.

So how can we, as beekeepers, intervene? Maintain healthy colonies by understanding their biology, providing good nutrition, biosecurity, and routine parasite control (varroa), and adequate shelter (ex. dry, well-ventilated hives). These basic principles apply to maintain the good health of all animals and humans. Strive to develop a keen diagnostic eye so you will know when to just monitor and when to "go in after the cork".

...When I returned the wine cork (still showing identifiable markings on the cork) to Red's pet parents, they were amazed because they had not drunk any wine for over a month. But this little fact further explained the case history, corresponding to the start of Red's stomach issues. Post-operatively, Red recovered nicely since we had discovered the ultimate root of the problem.

Euthanasia

Euthanasia: Our Bees Deserve the Consideration

One part of my job as a veterinarian has been to provide euthanasia services. It is not an easy part of our job, but it is a part of my veterinary oath to "relieve suffering". It is a service I am glad that I can provide to clients' animals when deemed necessary. Over the years, I have received many thank you cards, flowers, and other gifts from appreciative people who trusted me to guide the passing of their beloved animal. It is time we consider euthanasia and how it may be applied to our honey bees.

I am sure some of you have had to face the difficult situation of making the decision to terminate a hive. In the field of apiculture, standard methodology for honey bee euthanasia is still in its early stages, but I will outline the major caveats here.

Bees may need our help to best ends

The Definition of Euthanasia

The term euthanasia means "good death". According to the American Veterinary Medical Association (AVMA), "The term is usually used to describe ending the life of an individual animal in a way that minimizes or eliminates pain and distress. A good death is tantamount to the humane termination of an animal's life." (1) Periodically, the AVMA publishes guidelines that outline acceptable methods of euthanasia that veterinarians should use specifically and for a wide variety of animals. This 100+ page document contains some verbiage on invertebrate euthanasia, but a section specific to honey bees does not yet exist. However, I am on an AVMA committee that is currently working on such a project.

The AVMA says the following regarding invertebrate species (which would include honey bees), "While there is ongoing debate about invertebrates' abilities to perceive pain or otherwise experience compromised welfare, the Guidelines assume that a conservative and humane approach to the care of any creature is warranted and expected by society. Consequently, euthanasia methods should be used that minimize the potential for pain or distress." (1) The AVMA also recommends that each euthanasia technique for any animal consider the following criteria:

1. Ability to induce loss of consciousness and death with a minimum of pain and distress
2. Time required to induce loss of consciousness
3. Reliability
4. Safety of personnel
5. Irreversibility
6. Compatibility with intended animal use and purpose
7. Documented emotional effect on observers or operators
8. Compatibility with subsequent evaluation, examination, or use of tissue
9. Drug availability and human abuse potential
10. Compatibility with species, age, and health status
11. Ability to maintain equipment in proper working order
12. Safety for predators or scavengers should the animal's remains be consumed
13. Legal requirements, and
14. Environmental impacts of the method or disposition of the animal's remains." (1)

A few of the criteria above may not apply to honey bee colonies, but most do and could serve as a checklist for beekeepers and veterinarians to consider before euthanizing a honey bee colony.

How Veterinarians Use Euthanasia

In traditional veterinary practice, most veterinarians will agree to euthanize an animal for one of two major reasons:

1. A serious or terminal condition or illness that causes the animal suffering that cannot be reasonably remedied, or

2. An aggressive animal that poses a threat to people and/or the public. This is much the same for honey bee colonies.

Why We Euthanize Honey Bee Colonies

Typically, we consider euthanizing hives when they are severely ill, severely aggressive, and/or pose a public health threat. For example, in most states, hives diagnosed with American Foulbrood must be destroyed or it is highly recommended that they are destroyed, because they are most likely terminally ill themselves and can spread the disease to other hives. Severely aggressive hives may also be considered for euthanasia as they may pose a public health threat and may transfer aggressive genetic traits to future bee generations.

"Weak" hives may be a bit of a different story and a somewhat unique situation to honey bees. If the disease or condition in a "weak hive" is not severe or highly transmissible, we may be able to save a portion of colonies' bees by combining colonies and/or requeening. "Pinching" (and replacing) a failing queen may be a physical method of euthanasia for her and her genetics, but the colony may persist.

Some Current Methods:

There are several methods "out there" used for destroying a colony. I will briefly go through each method and point out their pros and cons. When euthanizing a colony be sure to have all your supplies ready before you start the process. It is best to start the process at dusk so all the bees are present in the hive, and few may escape the colony. Also close all the entrances to prevent escape as much as possible.

If you must burn the hive and colony after the euthanasia, make sure you are legally allowed to do so and you are following all local and state laws. If you must bury the burned hive remains, be sure to dig a pit big enough to accommodate the hive boxes.

Typically, this is at least 18 inches in depth. Don't forget to call 811 first. Then build a fire in the pit, so it is ready to receive the euthanized hive, if burning is necessary. Remember bees' wax is flammable.

1. Soapy water and/or vinegar/water solutions: These methods are fairly safe for beekeepers, but it does require opening the top of the hive to dump or spray the solution directly onto the bees through the frames. This may not be ideal for aggressive hives. It may also not kill all the bees quickly, especially if enough solution is not applied. Various mixtures have been used: ¼ dishwashing liquid to ¾ water, half and half water and vinegar with added soap, all with enough solution to soak the frames. If you must burn a hive euthanized by these methods, consider that a very wet hive may not burn as well. If you do not need to burn the hive and/or infectious disease was not the issue, the hive equipment can be cleaned and reused. Comb and honey should be discarded.

2. Isopropyl alcohol solution: If you have ever done an alcohol wash to check for mites, you will have noticed that alcohol kills bees quickly. For this reason, this method may be more humane for the bees. However, consider that isopropyl alcohol is flammable. You can dilute 70% isopropyl alcohol with water at least to a half and half solution. I have seen some claim that as low as a 5% solution will work, but I have not verified that myself. If you do not need to burn the hive and/or infectious disease was not the issue, the hive equipment can be cleaned, dried, and reused. Honey and comb should be discarded.

3. Fire: Again, if you are destroying a hive for AFB, fire is the preferred method. Dig the burning pit as close to the hive as possible to avoid moving the hive and possibly spreading spores. Be sure all entrances are closed. It is best to euthanize the bees first before burning the hive.

4. Professional chemical sprays: While utilizing pesticides is not generally recommended for killing honey bees and it will contaminate the environment and the hive equipment, if you are dealing with a particularly aggressive hive or a feral colony that has invaded a building structure, it may be time to recruit some professional extermination help. Always remember to consider your safety and the safety of those around you first.

5. Gasoline or Diesel: One recommended method of euthanizing a hive included pouring a small amount of gasoline or diesel fuel into the hive and then sealing the hive. The bees will die due to inhalation of the fumes. Bees should be checked for activity in about 15 minutes to see if a second application is needed. When using this method one must consider the flammability of these fuels, especially if you plan to burn the hive after euthanizing the bees. Diesel is less flammable than gasoline and would therefore be preferred. If you do not need to burn the hive, the hive equipment, comb and honey should all be discarded in a way that considers impact to the environment.

6. Other methods that are dangerous, impractical, and/or inefficient, and therefore not recommended: Gases (CO_2, ethylene oxide). Irradiation is rarely practical and can be costly. Some European resources point to the use of sulfur dioxide, however it is not readily available in the US and poses risks to an administrator.

The Best Ends

Because of the unique nature of our honey bees, they can present a challenge in performing a proper euthanasia. However, I believe we should take time to consider how we should best approach the situation and look to develop better methods of accomplishing a "good death" for our bees. More work should be done on this topic, and it would be a good opportunity for beekeepers and veterinarians to partner in coming up with better solutions. We should also take time to acknowledge and be empathetic to the emotional toll it may take on us, as beekeepers.

Immunity, Vaccines, & Honey Bees

As a veterinarian, I have administered thousands of various vaccines to a variety of animals over the years. As in humans, these vaccines are typically developed over decades of research and trials and are administered according to the individual's risk, including age, immunological status, potential exposure risk, economic feasibility, and geography. As a veterinarian, I have been vaccinated for a few more things than the average person due to my occupational risk. Rabies, for example. Yes, there is a human vaccine. Most veterinarians are vaccinated for Rabies because in our profession, we are likely to be exposed directly to this disease, a disease with nearly a 100% mortality rate. A nearly 100% mortality rate? Wow, why isn't everyone vaccinated?

The answer is three-fold:

1) Because we vaccinate our canines for Rabies. By controlling this zoonotic disease in the domestic dog population (which is the most common exposure risk for humans worldwide), we greatly eliminate exposure risk in humans. Unfortunately, in developing countries with little or no canine vaccine programs, about 60,000 people worldwide die each year of Rabies.

2) Because we have very effective therapeutics that can prevent disease after exposure, if applied early. Due to the longstanding, successful public education on Rabies, most folks know that if you are bitten by an animal, it should be examined for Rabies, and you should see your doctor asap. Even if you are bitten by an animal that is positive for Rabies, we have a combination of therapeutic vaccines and immunoglobulins (antibodies) that will prevent the virus from taking hold.

3) Because of #1 and # 2, the actual risk vs. benefit does not support the use of the vaccine in the general human population. Possible side effects of the vaccine outweigh the actual risk. On average, in the US, only 0-3 people die of Rabies annually, usually due to bat exposure. You are at least 9 times more likely to be killed by lightning (Check www.weather.gov).

Vaccines can be a wonderful tool in the control of infectious diseases, when applied appropriately.

Types of Immunity

To understand how vaccines work, it is important to understand how our immune systems work. I'll do my best to give you a brief synopsis.

There are three arms of defense in our immune systems, innate immunity, humoral immunity, and cellular immunity. Innate immunity is our first line of defense against any pathogen. Innate immunity includes the natural barriers of our body like our skin (or exoskeleton if you are a honey bee), respiratory and gut linings, natural microbiota, as well as non-specific reactions to any foreign invader, including certain inflammatory pathways, fever, enzyme activity, and our white blood cells recognizing and eating up non-specific, irregular stuff.

Humoral immunity largely involves antibody production that can help the body identify and eliminate pathogens from the body. Cellular immunity largely involves special blood cells called T-cells that recognize infected cells and can target the cells and/or pathogens within them. Both humoral and cellular immune systems have the ability to "remember" previously encountered pathogens and are therefore much more effective in eliminating the specific pathogen in any future encounters. "Immunity" has been achieved!

How we obtain immunity can also vary. We classify immunity obtainments as natural or acquired. Natural immunity can be passive or active. Passive, natural immunity comes from our mommas, usually passed through the placenta or milk. Passive natural immunity helps to protect fragile newborns from the rough and tough world, but it may not last very long. Conversely, active, natural immunity, we earn all on our own. We "naturally get" the disease. When infected with a pathogen we may or may not become ill and assuming we survive, our bodies typically develop immunity. Active, natural immunity usually provides a robust and often long-lasting immune response, but in some cases, there is a pesky risk of death, particularly with immune-compromised patients and/or diseases with high mortality rates.

Acquired means something outside of yourself has been given to you to help develop immunity. Acquired immunity can also be passive or active. Acquired, passive immunity occurs typically when a patient is given pre-formed antibodies as a therapeutic to help fight off a disease. Acquired, active immunity occurs when a patient is given a vaccine and develops immunity to the particular pathogen.

How Vaccines Work & Types of Vaccines

Vaccines are typically administered using oral, injectable, or nasal routes. Most vaccines have been developed for viral diseases, but we also can develop vaccines for some bacterial and parasitic infections. Vaccines can be used as a tool to accomplish three things:

1. Eradication of disease. This means there is no active disease left in the population on Earth! This is an extremely rare accomplishment, which has only occurred twice in the history of man or beast out of the thousands of known diseases that inflict us. Most people would guess and guess correctly that small pox is one of them. But the second one is a bit tougher.... It's Rinderpest...tricky, huh? Rinderpest is/was a cattle disease that was absolutely devastating to the cattle population particularly in Africa and therefore devastating to the human food supply and economy. What's next? Well... we're still working on Polio and maybe the Guinea Worm.

2. Elimination of a disease. Elimination means a previously existing disease is no longer present in a population in a certain geographical area, but it's still present in other parts of the World. Examples of diseases eliminated from the United States include Yellow Fever, Polio, and Malaria. This does not mean that the disease cannot re-emerge in the area if precautionary measures are ignored.

3. Control of disease. Control of a disease means that the disease is still present in a population, but it is reduced and manageable within the health care system, has a relatively low mortality rate, and/or has become endemic. This is the typical expectation and usually what happens with most diseases and vaccine use.

Vaccine forms have varied and evolved over the years. "Live" vaccines are typically "attenuated", which means the vaccine contains a weakened form of the actual pathogen. Attenuated live vaccines have many advantages in that they provide a robust immune response of both humoral and cellular immunity, but since the vaccine is still a form of the pathogen, there is a low risk of the disease developing, therefore, live vaccines are not recommended to be given to immunocompromised patients. Examples of live attenuated vaccines include flu vaccines, MMR, polio, smallpox, and Chicken pox/Shingles.

Some vaccines are in the "killed" or "inactivated" form. These vaccines may induce a less robust immune response but are safer since the pathogen is killed and there is no risk of disease development. Killed vaccines are recommended in immunocompromised patients and with pathogens that are extremely virulent. For example, Rabies is a killed vaccine, as is the vaccine in development for American Foulbrood (AFB). For good reason!

Even with rigorous trials, vaccines are not always perfect. All vaccines can have side effects. Sometimes they are taken off the market for one reason or another, including safety concerns, lack of efficacy, and economics, or better, safer vaccines are developed to replace them. The risk vs. benefit of a vaccine must be weighed with a consult from a patient's doctor.

Immunity in Honey Bees

Because honey bees do not produce antibodies, we used to think complex individual immunity, particularly humoral immunity did not exist in honey bees, but the more research we do, the more complex we find honey bee immunity to be. We now know that individual immunity exists in honey bees and includes a mix of complex innate, humoral, and cellular responses.

Social immunity is also a characteristic of the honey bee colony. Hygienic behavior, propolis use, and thermal elevations are examples of social immunity in action. We know that individual sick bees will fly off and die in hopes of not infecting the rest of their siblings. Nestmates share and transfer just about everything, so feeding worker bees a medicine easily distributes it throughout the colony.

The Vaccination Process in Honey Bees

The vaccine/s under development are/will be an oral vaccine, fed directly to workers and then ultimately to the queen through royal jelly. Over years of research, a company called DALAN, together with other academic researchers have uncovered the importance of **trans-generational immune priming** in honey bees. This is a process, often seen in insects, where the mother can pass immunity to previously encountered pathogens to her offspring. The process involves innate and acquired immune activity in the queen those results in something analogous to passive transfer in mammals. Through various immunological techniques, researchers are able to tag, often using a fluorescent signature, the vaccine (killed pathogen particles) to see how they move through the queens' bodies to their offspring.

Results of these studies showed that larvae from vaccinated colonies showed a significantly higher survival rate than larvae from unvaccinated colonies. Over the years, DALAN researchers with their academic colleagues also conducted comparison tests for negative effects on colony strength, honey quality, honey quantity, and queen fitness. So far, these studies revealed no negative effects associated with vaccine use which supports safe vaccine use within a hive.

What's on the Horizon?

There are a few questions I am sure beekeepers will be asking about AFB and other honey bee vaccines. When can we expect them on the market? How much will they cost? What about practicality of use?

In 2023, DALAN has released the vaccine to the American market for the first time with emergency approval. It will be available in Canada starting in 2024. Beekeepers are able to learn more about the vaccine and make a purchase on DALAN's website. As far as practicality, the company anticipates that the vaccine will likely last a season and will need to be applied each beekeeping season. If the benefits of these vaccines are to increase honey bee immune resistance to AFB, EFB, honey bee viruses and fungal diseases, and they eliminate antibiotic over use/misuse/resistance, it will be priceless to our honey bees, the beekeeping industry, and public health in general.

Frame Seven - Honey Bees Diseases

Like any other animal, honey bees are subject to a myriad of various types of diseases, bacterial, viral, fungal, parasitic, nutritional, or traumatic. This frame features a few unique maladies of honey bees and the approaches we take in an attempt to prevent or fight these nemeses.

A Spunky Drone & PPE

Tropilaelosis

Zoonotic Disease and Public Health is a course I have had the joy of teaching to pre-health students for over 10 years. The course's content covers all types of infectious diseases and how they affect **and** connect humans, animals, and the environment. Some of my students affectionately call it "Zoo" or even the "poo" class, because so many diseases can be transferred through fecal/oral contamination. Eww, but true! Many students are amazed to discover how many diseases surround us, how the health of our world is so interrelated, and under normal circumstances, how most of us are blessed with a wonderful immune system. Certainly, these same observations apply to honey bees and their health.

Emerging infectious diseases is one topic we cover during the "zoo" course. Emerging infectious diseases are infections that have recently appeared in a population of humans or animals. Emerging diseases often arise when they are brought into new geographical ranges and/or species. Some causes of emerging diseases may not have been previously known, while others may already be known, and pose a serious threat, if they are able to increase their geographic range. Ebola, Zika, Rocky Mountain Spotted Fever, Varroosis, and COVID-19 are all examples of emerging diseases. Many emerging diseases often originate from "foreign" or "exotic" diseases (or newly named "transboundary diseases"). Foreign, exotic or transboundary diseases are diseases that naturally exist in a certain country, continent, or areas of the world, but may cross borders, continents and/or oceans to infect new regions. If allowed to move into new geographical areas, foreign diseases can emerge in a population with little natural immunity against the disease agent. Therefore, these diseases can cause high morbidity and/or mortality when introduced to the new population of animals or humans. In our modern world, international trade and travel often accommodates hitch-hiking diseases and pests. To safeguard animal health in the US, a list of foreign animal diseases (FAD) is continuously monitored by the USDA and accredited veterinarians.

But what about bees? Do they have a current "FAD"? They do. It is a parasitic disease of honey bees that does not always make the headlines, but mirrors examples of other disease processes we see highlighted in our world. No, it is not the "murder" or Asian hornet, but a disease that is and should be on beekeepers', entomologists', and veterinarians' radar: Tropilaelosis.

Tropilaelosis is a mite infestation of *Apis mellifera (European honey bee)* caused primarily by two major species: *Tropilaelaps clareae* or *Tropilaelaps mercedesae*. These mites' natural honey bee hosts (*Apis dorsata, Apis laboriosia,* and *Apis breviligula,* "giant" honey bees) are better adapted host species of honey bees compared to *Apis mellifera*. Their natural range is found in Asia, Indonesia, and the Philippines. The mites have also been reported in parts of Africa, including Kenya and the Republic of the Congo. Tropilaelosis is currently a disease regulated world-wide and monitored by the OIE (The World Health Organization of Animals) as a notifiable disease and the USDA as a reportable disease. These mites are one reason why honey bee importation is limited in the US.

Be on guard for emerging diseases

The lifecycle of the mite is somewhat like *Varroa* with the reproductive cycle involving a gravid foundress mite invading a brood cell, egg laying, developing mites parasitizing and often killing the larvae/pupae, and re-emergence of new adult mites. Compared to *Varroa*, the reproductive cycle is relatively short, only about one week, and all mites emerge from the brood cell including the males. This feature allows the *Tropilaelaps* mites to populate a colony much faster than *Varroa* and therefore, take down a colony quickly. *Tropilaelaps* mites are unable to feed on adult bees, so their phoretic phase is much shorter than *Varroa*, usually only three days. This characteristic force the mites back into the brood for yet another quick reproductive cycle, killing more brood and making more mites. Despite the short phoretic period, adult bees are still able to spread mites to other hives via swarms, package bees, exchange of frames bees between hives, drifting, and robbing.

Mites are diagnosed and treated using similar methods to *Varroa*. Adult honeybee samples can be checked for mites with alcohol wash or sugar roll. While mite counts levels have yet to be established for *Tropilaelaps*, any mites found would be significant. The mites are visible with the naked eye, but they are smaller and move faster than *Varroa*. They are easier to observe in capped drone brood. Sticky board or "bumping" frames to dislodge mites onto a white surface can also be used for detection. At the colony level, infestations will result in rapid colony collapse or absconding. Brood comb may be severely affected due to high mortality infected larvae and pupae. Treatment should involve an IPM approach. Treatments can include common acaricides used for *Varroa*, along with biological controls of inducing brood breaks, brood removal and caging the queen. Treatment timing protocols should consider the short phoretic period of the mites. Being unable to parasitize adult bees is one biological weakness of *Tropilaelaps*, that we can exploit. Natural brood-less periods and overwintering are ways to limit or control these parasites. Luckily and so far, the geographic range of *Tropilaelaps* has largely been limited due to this "tropical nature" of the mite. However, some honey bee colonies in South Korea, with a more temperate climate, have been found to support *Tropilaelaps* mites.

There is more bad news. While rare, *Varroa* and *Tropilaelaps* can co-infect colonies, but *Tropilaelaps* usually out competes *Varroa*. *Tropilaelaps* has also been found to be a vector for viruses, like DWV.

The good news: Tropilaelaps has not yet been reported in much of the world, including the US, Europe, Australia, and Canada. However, awareness and prevention of diseases are keys to keeping our honey bee population safe. How diverse animal species, humans, and diseases can be, yet how much is still shared and interconnected, amazes me. Studying and understanding these similar biological and epidemiological principles are paramount to understanding how we can all work together to best promote our collective health.

Hive beetle larvae

Not as Fun as a Volkswagen

Winter is a time where we see breaks in lifecycles of many parasites that can play to our advantage. Many parasites diminish in number and/or reproductive activity or hibernate during the winter months. This does not mean that they are gone, however. If given the opportunity, parasites of all types will hunker down in/on our warm and cozy bodies, home, or hives to make through the winter and produce the next generation.

Most parasites, including fleas, ticks, Varroa mites, and various intestinal worms are endemic in most of the world. We cannot stop them but strive to control them. No matter the type of parasite or animal it infects, several factors are always at play in the host-parasite-environment relationship. Nutrition, general health of the host animal, genetics, medical preventative or treatment interventions, and climate/shelter/other environmental stressors all play a role in maximizing or minimizing a parasite's effects on our animals.

Nope- this is not another article on Varroa mites. By reader request, I will review a parasite that is often listed as a secondary "pest", but they can still cause significant issues within a hive and may be an indication of the even greater issue of a weakened hive...*Aethina tumida*, the small hive beetle.

History

Aethina tumida is a newer pest to the US, arriving as early as 1996, about a decade after we first observed that Varroa had arrived on our shores. In Europe, it is considered a notifiable pest. Small hive beetles are native to South Africa, but despite efforts to control spread, hive beetle infestations are now found throughout the US, Canada, Mexico, Australia, and parts of Europe, Asia, South America, and Central America.

Adult small hive beetles are small (about .5cm), brown to black, oval-shaped beetles with club shaped antenna. Female beetles lay their eggs in the hive. Their eggs are smaller than honey bee eggs, so they are hard to individually see, but the eggs may be laid in cells or in multi-clusters around the hive.

The larvae are creamy white to light yellow in color and can be over 1 cm long. Larvae feed on all things within comb; honey, pollen, and brood, and can be quite destructive. Pollen patties may also attract female hive beetles for egg laying and therefore may become infested with larvae. Adults and larvae can defecate through the hive, ruining honey by fermentation.

The pupae must mature in soil, typically in front or near an infested hive. After pupation, the adults emerge, are sexually mature within a week, and may reinfest the nearby hive/s or travel up to five miles to find a new host. The lifecycle from egg to adult is typically 4-6 weeks. Adult beetles may live up to 6 months and females can lay up to 1000 eggs in her lifetime. Warm and humid conditions favor hive beetle reproduction.

Aethina tumida are "timid" in their behavior in that they use hiding as a survival behavior. Adults move very quickly and can be elusive to both the bees and the beekeepers. In some hives, bees have been observed "herding" beetles into an area or propolis "jail" to control them by confining them.

While parasitic to the honey bee, A. tumida have been found to infest bumblebee colonies, and to feed on tropical fruits. Small hive beetles have been reported to be a possible vector for *Paenibacillus larvae* (American Foulbrood) and some honey bee viruses, like deformed wing virus (DWV) and Sacbrood virus.

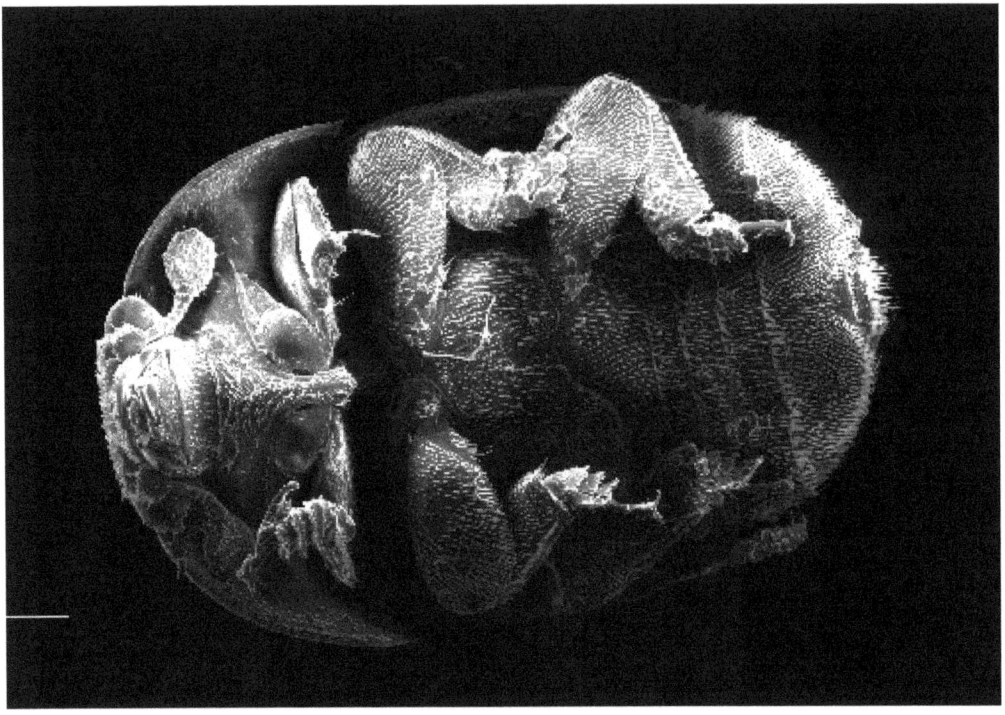

Scanning Elecrron Micrograph of a Hive Beetle

Diagnosis/Clinical signs

Diagnosis can be straight forward with visualization of adults, larvae and/or eggs. Soil in front of hives can be examined for pupal tunneling. However, several non-pathogenic beetle species, other than A. tumida, can be found in hives. So, it is vital that beetle samples are properly identified. Visual identification can be aided with laboratory microscopic exams and PCR is available for confirmative diagnosis.

Damage to the hive's comb may be evident on inspection and a foul odor due to fermenting honey may be present. Slime trails may be evident from larval wanderings. In-hive beetle traps can be used for monitoring as well as control (see Traps below).

Hive beetle infestation should also be on the differential diagnosis list for hives that abscond.

Prevention/Treatment/Control

Chemicals

Many who comment in the literature are not big fans of using insecticidal products around bees to control hive beetles. I will say that I am in that camp. However, to be complete, there are several products on the market that have been used in the control of hive beetles. I will briefly mention them here.

Coumaphos is an organophosphate that has been used in veterinary medicine starting in the late 1950's for the treatment of various parasites in a variety of animals. While attending veterinary school in the mid 1990's, I mostly learned about organophosphates as toxins. Coumaphos persists in the environment, has shown resistance in the treatment of target parasites, and many of the long-term effects of chronic human exposure remain unknown (CDC). Hives are treated with coumaphos strips placed **directly** into the hive.

There are several ground drenches used (i.e., not placed in the hive but on the ground in front of the hive), which contain Permethrin. Permethrins have lethal and repellant effects on various external parasites and pests of many animals and humans and is commonly used medically today. It can be helpful as a medication, when used under proper guidance, but it can persist in the environment. It is particularly toxic to cats and aquatic life. Do not use it around bodies of water, if you like to fish. Like other insecticides, it is toxic to honey bees, if they come into contact with it.

Gravel in the bee yard may prevent pupation of Hive Beetles

Traps & DE

I would be happier if beekeepers utilized fewer toxic methods, for both us and the bees, to help monitor and control hive beetles. There are several beetle traps and beetle towels with or with the use of oil that are placed in the hive on the market. Homemade traps can be made from dry, disposable, dusting mop pads. These traps and towels can help trap adult beetles so they can be removed from the hive and may I suggest, placed in a hive to measure the presence or amount of hive beetles in a hive. Remember you will have to check your traps periodically for them to be the most informative and effective. Anecdotally, I have heard varying reviews from beekeepers regarding their experience with various trapping methods utilized in hive beetle control.

Diatomaceous earth (DE) makes me recall my old lifeguarding and swimming pool "work" days. DE is commonly used in filtration systems of swimming pools. Diatomaceous earth originates from the remnant shells of little pre-historic sea creatures called diatoms. Evidence of DE's effectiveness in controlling hive beetles is anecdotal, but the theory is that the crystal-like shells of diatoms cut and dehydrate any hive beetle larvae that try to pupate in front of a hive. Therefore, this organic substance can be placed outside, in front of hives, to make the ground less hospitable for pupating hive beetle larvae without the fear of environmental toxicity. Check your local swimming pool store for prices. Warning: DE can become very clumpy and stick to your boots when it gets wet.

General management/IPM

There are several management techniques that can also be employed to reduce hive beetle infestations.

1. This is often stated but cannot be overlooked here- **keep strong colonies**. Strong colonies will always have an advantage over parasite infection and hive collapse. Colonies collapsing due to hive beetles may have had underlying issues in functional hive immunity. Assure varroa control and good nutrition.

2. This may seem obvious but physically remove and kill any hive beetles you see during hive inspections. Death by hive tool. Do not spend all day doing this but any beetle you can take out can be helpful. Be aware they are quick little buggers.

3. This may not be practical for all beekeepers, but it is a method I have employed with success. Do not keep your hives directly on the ground/dirt. Placing a weed barrier and then gravel over the ground will create a barrier for grass, weeds, and hive beetle reproduction. I suspect hives on rooftops may enjoy a similar benefit.

4. While it's true that hive beetles can "fly in" to your hive from miles away, biosecurity principals should still be employed to reduce parasitic pressures. Swarms, while they may have other issues, are less likely to introduce hive beetles to a new yard. The trade and movement of package bees, nucs, colonies, various hive products and even tropical fruits can present risks of further beetle spread. If you are purchasing nucs or hives, ask the supplier if you can inspect the hive/nuc, prior to purchase. If the answer is "no" that may provide you with enough information to make an informed choice. Nucs originating from the South are more likely to carry hive beetles, since the winter is mild in those areas and their reproduction is not curtailed (Sammataro). Producing your own new stock from splits is often the best way to avoid biosecurity breeches.

5. Freezing frames will kill any hive beetle eggs (as well have other parasites like wax moths). If possible, store un-used comb over the winter in a freezer.

6. Hive beetles do not prosper or reproduce if humidity falls below 50%. Controlling humidity within hives can reduce their success. Top entrances and screened bottom boards may be helpful under the correct environmental conditions.

7. In honey houses, extract honey from supers quickly to avoid attracting hive beetles.

What's on the horizon?

The Large Hive Beetle, consisting of two main species that infest honey bee colonies, *Oplostomus fuligineus* and *Oplostomus haroldi*, are currently found in southern Africa. This pest is **not currently** in the US, but it is just another parasite to be on our watch list.

The Diagnosis of the Shrew

One thing about beekeeping is it always provides an opportunity to learn something new. It definitely keeps one humble trying to keep up with new (while remembering the old) information. In a recent conversation with a few Honey Bee Veterinary Consortium colleagues, one member was trying to figure out a sudden early Spring collapse of a single hive located in Iowa. The main evidence was a large pile of dead bees that looked chewed up with the thoraxes apparently hollowed out (see photos provided). Long story short, the conclusion was made that pygmy shrews were the main culprit. Pygmy shrews...hmmm. I vaguely remember reading about them as a pest in Canada, but they really were not on my radar as a common differential for a cause of a weak colony coming out of winter and/or colony collapse. Turns out they are a common pest to add to the list in the US, particularly northern States. In my research of this pesky mammal, I found that many beekeepers may be unaware of this pest, so I thought I would devote an article to help bring more awareness to this little creature.

Pygmy shrew (*Sorex hoyi*) fast facts

- Common names used: Northern or American pygmy shrew.
- *Sorex minutus* is not the same shrew but found in Europe and Asia.
- Very small, smallest American mammal by weight with an average of 3g.
- Can fit through a hole less than one cm- so typical mouse guards will not keep them out.
- Pointy snout with whiskers.
- Have a tail almost as long as their 1 1/2-inch body.
- Grey in color.
- Eats primarily insects and arachnids.
- Take bees from the outside of the cluster and then consume the bulk of the thorax containing the delicious meatball of flight muscles.
- Fast moving, so they are hard to catch a glimpse of...

Clinical Findings in a Colony

- Pile of dead bees, bodies macerated, thoraxes eaten or hollowed out.
- A more common reason for decapitated bees. (Unless you are in Washington state area, it is very unlikely Asian hornets are the cause of decapitation).
- Shrew droppings may be found -they look similar but are somewhat different

compared to mice droppings.
- Finding a dead shrew in a hive is the only 100% way to confirm a diagnosis.
- Shrews do not eat honey/pollen. Stores will be left undisturbed unlike mice infestations.
- Pygmy shrews can be a cause of colony weakness or colony collapse. There may or may not be signs of other diseases.
- Seasonality: active bees keep shrews away, so shrews are a winter, early spring, issue.

Typically, only one or a few hives are affected. This may not be important for a larger operation but if you only have one backyard hive the results could be devastating. The literature reports that in Canada the shrews have adapted to invading hives in the winter as a survival strategy, so some Canadian commercial operations unfortunately have had considerable losses due to pygmy shrews.

Prevention or Remedies?

Assessing your risk for pygmy shrew invasion is up to you, whether it is worth it for you and your operation to apply preventative remedies or not. Wire mesh can be applied going into winter at hive entrances. Recommendations range from one quarter to three eighths of an inch. Such small entrances are harder for bees to get into the hive especially with pollen loads, but the three eighths of an inch mesh may be more forgiving for honey bees to enter the hive with pollen. Beekeepers may choose to apply the mesh only in the late Fall and remove it in the Spring. Again, it is a risk versus labor cost assessment for beekeepers to consider but it is something else to keep an eye out for now that you know the signs.

Emerging Diseases in Honey Bees

As a Pennsylvania girl, I grew up with Rabies. Clearly, I do not mean I had the disease rabies, but that this deadly disease was all around me - endemic in our mammalian wildlife population. We, Pennsylvanians, knew how to handle it. We took our dogs and the cats we could catch to the Rabies clinic, and at a young age, we were pulled aside for "the talk". The talk went something like this: "If you see a weird raccoon or groundhog in the yard, do not touch it, but call mom or dad to come shoot it." End of talk and end of report. You can imagine my surprise (and several of my PA classmates' surprise) when during our freshmen year in veterinary school, Ohio was freaking out about Rabies, like it was a new thing. Well, it was new to them. In the mid 1990's Ohio had their first reported case of Rabies in terrestrial (non-bat) wildlife. Us Pennsylvanians just blinked, shrugged and were like, "Wow, they didn't have Rabies in the State, at all? I suppose they'll have to learn the talk." To this day, government programs distribute rabies vaccine baits by throwing them out of airplanes and trucks along the Ohio Pennsylvania border, in hopes of vaccinating rogue raccoons that may enter the State from the Commonwealth. However, Rabies is now considered endemic in wildlife populations in Eastern Ohio.

There are countless case studies and lessons about the emergence and spread of different diseases in humans and animals throughout the World over time. Honey bees are no exception. Consider what is currently going on in Australia with emerging parasites, Varroa and the Braula fly. We naturally want to do anything we can do to eradicate diseases from the face of the Earth, however actual eradication is almost never a true reality. In previous chapters (Tropilaelosis & Immunity, Vaccines & Honey Bees), I defined emerging diseases and explained three levels of disease management.

Considerations in population medicine

Population medicine or herd medicine are terms that we use to describe the concepts of looking at disease management from a group perspective. Sometimes this group is a single herd, yard, or flock of animals, sometimes it could be the entire human population of a country or even the World. There are certain fundamental principles that should be applied in population medicine challenges.

1. **Understand the methods of spread of the disease:** To effectively develop controls for the spread of disease, we must fully understand the way/s the disease is transmitted. For example: Managing sexually transmitted diseases versus managing aerosol transmitted diseases would demand different protocols and recommendations.

2. **Understanding origin and scope:** The origin (geography and species) of a disease can clarify the natural epidemiology of the disease, so we may be better able to recognize symptoms, transmission methods, and expected morbidity and mortality of the disease. Knowing the existing geographical scope and incidence within human or animal populations is important to evaluate, as managing isolated verses global cases are very different.

3. **Understand we are limited by our diagnostics:** Surveillance testing is a key component to monitoring possible emerging diseases or changes in diseases' incidence and is appropriate before and at the beginning of an outbreak. If you do not test for something you will not find it, but if you do test for something you are likely to find it... sooner or later. This can be a double-edged sword. How much information is necessary to switch surveillance to management? "Contact-tracing" may be helpful in the initial stages of an outbreak but over time, the lines on the map just merge into one big blob.

   No testing methods are perfect. All diagnostics are subject to sensitivity (positive results are truly positive cases) and specificity (negative results are truly negative cases) percentages. We also know that the "first" positive case we find of a disease in an area is actually an indication that the disease is already there...maybe for a while.

4. **Geography and weather may play a role.** In considering geography, islands often have the unique benefit of isolation that can make disease-free areas more possible. Many islands, like Australia and Hawaii, may have very strict biosecurity laws at ports of entry to keep diseases at bay for good reason. After the land is breeched, however, oceans no longer serve as a barrier. With the globalization of our world, natural and man-made geographical barriers to disease are becoming less and less effective in keeping pathogens in-place.

   Natural weather patterns can encourage disease emergence or not. For example, many diseases thrive in warm, humid conditions, while cold winters may limit the scope of a disease.

5. **Does it matter?** Sometimes finding something may be incidental. Is it worth doing something about it? For example, during my tick studies we identified a tick species that had not been clearly identified (at least in the official literature) in Pennsylvania before. This was an interesting finding but of little significance to our study because the tick we found is not considered a vector of disease in humans.

6. **Is it even possible to eliminate?** We know that true "eradication" is almost

impossible for diseases in general. Elimination has been achieved before for certain diseases, but it often requires a non or mildly contagious disease with the employment of intense measures, including, isolation of infected populations, culling of (animal) infected populations, and highly effective vaccinations. Expecting the disease to become endemic may be the most realistic and best hope. Expecting, equipping, and employing "control" measures early within a vulnerable population may lessen the impact of the disease as it moves through a population from an epidemic to endemic stage.

7. **Above all else do no harm**. The "cure" should never be worse than the disease. The morbidity (rate of illness) and mortality (rate of death) should be considered when applying disease management protocols and making recommendations. Social, emotional, and economic impacts of a disease response should also be weighed in any decision making.

Recent Australian emergence examples: Varroa and Braula fly

My heart goes out to Australian beekeepers who attempted to stop the spread of Varroa mites, which were detected on the island continent for the first time at the Port of New Castle, New South Wales in June 2022. Australia has admirable biosecurity guidelines for honey bees and up until this point, have amazingly, enjoyed a Varroa-free industry. However, I am afraid the Australians are about to join the rest of the world and will have to learn to manage varroa mite infestations within their hives.

In 2022-23, the Australian government tried to prevent further spread of the mites into the country by issuing lockdowns of hives, restricting sale of honey, tracing possible contacts, and euthanizing/burning of all hives within designated and increasing geographical radiuses. The result was the loss of thousands of hives, millions in economic losses in hive, honey, and pollination resources, conflict within the industry, and devastating emotional toll on bee farmers, all with the Australian spring just starting. Despite these efforts, each of the latest news reports I read only conveyed further spread. Given the facts that Varroa has made land fall in Australia, mites reproduce exponentially, honey bees fly and swarm, and the history of Varroa spread around the globe, I did not see elimination as a realistic outcome. Time for the employment of the control and management phase of Varroa mites has come to the continent.

In the US, we know that Varroa is a major **contributor** to the 40%-45% annual loss of our hives. However, a closer look at the data will show that commercial beekeepers experience at least half the annual loss of hives compared to the average of all beekeepers.

Given that burning hives has a 100% mortality and 100% economic loss, one must consider that accepting Varroa as a portion of a 20-30% annual loss, at some point, becomes a better alternative.

Here's some good news for Australian beekeepers. The rest of us have been managing Varroa for decades, we have learned a lot and have many tools in the toolbox. It has not been easy, but American and European (largely commercial) beekeepers have maintained our total colony numbers over the last several decades despite Varroa. Our beekeepers and honey bees are still able to support the top agricultural and honey producing countries in the World. We can and should come alongside Australian beekeepers with empathy in learning how to detect, manage, and treat Varroa mites within their colonies. I believe the Australian government is aiding beekeepers with financial support for their losses. Mental and emotional health support should also be part of the recovery plan.

Another interesting development is the additional new finding of the Braula fly during surveillance for Varroa mites in Victoria. Remember, when you look for something, you may find it and maybe even find something else. While Braula is certainly less of a threat to honey bees than Varroa, they can damage honey and comb. Because Braula tends to hang out on the queen, this wingless fly could be transmitted through queen trade.

What's next?

I wish I could tell you there are no more diseases that will emerge in honey bees. The Asian giant hornets, Large hive beetles, Tropilaelaps are all on the horizon. My best advice...Do not give up on biosecurity. You may not be able to control what everybody else does or everything the bees do, but you can at least control what you bring in and out of your own yard. And remember, do not play with weird raccoons.

Always consider biosecurity

Frame Eight - Honey Bees, Humans, and Public Health

There is no doubt honey bees are adored by humans. We love them for their honey, beeswax, and hardworking yet whimsical nature. Images of honey bees adorn everything from jewelry to pillows to kitchen wares. Beekeeping is a romantic endeavor to be sure. The relationships we share with our bees and other bee enthusiasts create a community that can bolster our social, physical, and spiritual health. This frame focuses on several examples of those relationships from the perspectives of multiple types of bee people. Honey bees are at the center of various public health issues, including food security, proper use of antibiotics, and environmental health. Beekeepers should be aware that their interactions with their bees can have local, regional, and even global impact on the health of every one of us. Here are a few examples of public health implications in apiculture.

Our connections with Honey Bees

"Good" Help

On a rare day off in mid-May, I went fishing with my family, ironically at Slippery Rock Creek. I found a beautiful tributary off the main stream and walked up the bank to a deep hole I found amongst the car-sized rocks. I had my eye on reaching a large flat rock in the middle of the stream, to plop down and go after a few trout. While traversing a few steps to get to my prized destination, my feet failed me, I slipped, wedged my knee between a rock and log and fell, tearing my anterior cruciate ligament (ACL) and smashing the tibia of my right leg. Even before the doctors' visits, X-rays, and the MRI, I knew instantly something was seriously wrong. One of my first thoughts was what am I going to do now? How am I going to work? Take care of my family? Build my house? Take care of my bees? Perhaps other beekeepers can relate to "What if I get hurt? How am I going to do life?" I certainly know several Beeks that have struggled through managing their yards with health problems and injuries.

Here is the good news... While surgery for an ACL repair can put you on your back for several weeks followed by months of rehab, my orthopedic doctor did not recommend surgery for me but instead said I was a good candidate for physical therapy rehabilitation. He said that in his experience and depending on the individual patient, ACL reconstructive surgery is often a mistake and may cause more harm than good. As a veterinarian, this concept made complete sense to me, but as a patient, I needed to hear it from my physician.

Benign Neglect

In veterinary and human medicine, we have a treatment called "benign neglect". It essentially means that as a medical professional, after evaluating a patient with a condition, we determine that the best course of action is to do really nothing (with perhaps just supportive care) and wait. Time and the amazing body often heals itself, if given the chance. "Above all else do no harm", "Go home, rest and drink plenty of fluids", "Take two aspirin and call me in the morning", and the placebo effect are all common illustrations of this approach. When diagnostics or treatments *are* started, we begin with the least or minimally invasive approach. For example, we may try medicines or therapy first, before surgery to address a certain condition. Less can be more in certain situations.

I know that I have preached before in previous chapters that it is essential for beekeepers to learn how to do a complete hive assessment or exam. That is still very true, however once you have learned how to do a complete hive inspection, you should also be learning when to stop. Risk assessment, risk vs. benefits, should be involved in everything you do in a hive. The why and the how of what you plan to do before opening a hive should be well defined. For example, if you find white wax and eggs do you really need to go through every frame to find the queen to confirm queen status? ...maybe, maybe not. You can often gain information needed just by observing the exterior of the hive and entrance, just by "popping the top" and looking in, or simply pulling a few frames. The necessary extent of hive inspections may vary by season, hive/yard history, and the type of beekeeping operation. "Surgery" is not always needed. "Everything in moderation" applies to our invasive behaviors when working our hives, but it comes down to a (hopefully educated) judgement call by the beekeeper. As a beekeeper you may need to recruit some help from a seasoned beekeeper or veterinarian on what level of intervention your hive/s may need.

Words as sweet as honey comb

Helping each other and how beekeepers can help vets...

Every one of us needs help sometimes. From what I have seen, most beekeepers are pretty good at forming groups that lean on each other in tough times, but I believe this symbiotic relationship can be expanded. I have listened to the concerns that beekeepers have about the difficulty of finding a competent, local veterinarian to help them. From the veterinary side I can share this: Currently, bee medicine curriculum in US veterinary schools is being developed and delivered, more honey bee veterinary textbooks and references are being created, hundreds of hours of honey bee continuing education for practicing veterinarians have been developed and delivered, and a bee vet certification program is under development. But how can beekeepers get involved on our end? Here are a few action steps beekeepers can take to help veterinarians.

1. Invite your local veterinarian/s to your bee yard to shadow you.

2. Invite your local veterinarian/s to your beekeeping meetings. Most may say "no", but you only need one in your area to take care of the local bees.

3. Reach out to local Veterinary Medical Associations (VMAs). Like beekeeping clubs, most states have state and/or regional veterinary associations. Most VMAs are interested in learning more about bees and working with beekeepers.

4. Reach out to your state Department of Agriculture (DA). Most states have apiarist/s and state veterinarians working at the DA. These groups have started more conversations lately and may have resources available for you or be interested in collaborations.

These help suggestions do not have to be as difficult as "surgery" but provide supportive care for a healthy, symbiotic relationship. As for me and my broken knee, luckily, I have "paid it forward" by previously allowing a neighborhood teenager to shadow me at my home bee yard and I have two trained summer research college students to help me do much of the "heavy lifting" around the College yard this summer.

Bees in Christendom, the Spiritual Health of Honey Bees

Poor mental and psychological health is an ever-increasing public health crisis in our country and world. As a college professor, I am on the front lines seeing this epidemic as it manifests in our young people. The statistics are appalling. Poor psychological and mental health even affects our animals. However, in talking with other beekeepers, one common theme I hear from them is the peace they receive when tending their bees. Recently, I have had the opportunity to explore perhaps one reason why. After discovering that honey and bees have over 60 references in the Bible, making them on par with camels...I decided to take a closer look. During a Fall semester, I signed up to create and deliver a 5-week chapel series at our College on "Honey and Bees in the Bible". In addition to the opportunity to teach the community some fun facts about honey bees, we also get a chance to see what the Almighty thinks of bees. Well, spoiler alert... He's a fan.

The series' presentations each include the following: Introduction to topic and outline, Fun bee facts/demo, Scriptural references and reflections, Relevance and application to our faith and walk, and a Question/comment period. Five major themes are covered, one per week. The themes are: 1) Community: The Fruits of Hive Work, 2) God's Provision: The Land of Milk, Honey & Health, 3) The Fall: Disease in the Garden, 4) God's Prized Possession: The Sweetness of God's Love for Us, and 5) Stewardship: Our Relationships with God's Creation. I am sure just about any beekeeper could easily see how honey bees could be used to demonstrate these themes.

Turns out there are two basic types of references to "bees" in the Bible. The first category is not a reference to honey bees or honey at all but to hornets or wasps. These Biblical verses represent God's judgement, protection, promise, and power. Generally, God getting after someone. Here's some references to check out: Exodus 23:28, Deuteronomy 1:44, Deuteronomy 7:20, Joshua 24:12, Psalms 118:12, Isaiah 7:18.

However, most verse references in the Bible to "bees" are about honey and/or honey bees. Honey and honey bees are used as images to convey God's blessings, God's love, strength, wisdom, things of great worth, and even references to Christ. Some Biblical names, like "Deborah", the name of the famous and only female leader of Israel in the time of the Judges, means "bee" in Hebrew.

I work with a company, AI Root, who produces beeswax candles especially for churches. The Catholic church and other churches require candles to be made primarily of honey bees' wax due to the purity, value, and holiness that only beeswax represents. Pretty cool, huh?

Since honey bee colonies demonstrate individual roles and purposes within a united community with more difficult tasks taken on through time, they are also a beautiful illustration of the Body of Christ and the sanctification process. Here are a few verse references: 1 Corinthians 12:25-27, Matthew 18:20, 20:26-28, 22:39-40, Galatians 6:2, Romans 12:5, John 15:12-13.

Honey bees demonstrate our relationship to the land and God's provision through crop pollination and hive products. Here are some references: Exodus 3:8 3:17, Deuteronomy 8:8, 11:9, 26:15, 27:3, Leviticus 20:24, Ezekiel 20:6, Jeremiah 11:5. Other references illustrate honey as a good gift or food, some examples, Proverbs 16:24, 24:13, Matthew 3:4 (Mark 1:6), Exodus 16:31. Honey is also considered to be of great value in the Bible, Ezekiel 27:17, 2 Chronicles 31:5, Genesis 43:11, as well as, a source of strength and wisdom,1 Samuel 14:27-29.

Certainly, everything is not always perfect in the garden. As beekeepers we face depressing losses due to a myriad of honey bee diseases and environmental decay, but the Bible addresses that, too. From the Fall in Genesis chapter 3 to the anticipated redemption of creation in Romans 8:18-22, beekeepers can take heart.

Even with loss there is still redemption and with redemption comes joy. Consider the dances of the honey bee when they have discovered a good thing. That honey, so carefully made, never really goes bad. Life can even emerge from death.

One of these redemption stories and probably the most famous bee story in the Bible is found in Judges 14, the story of Samson and the bee "swarm" in the lion carcass. Since becoming a beekeeper, I have a different take on the story than before I was a beekeeper. The bees in the lion are referred to as a "swarm", however the text goes on to explain that the bees had built comb within the carcass. So, beekeepers would know this was not just a swarm, but an established colony that may have come from somewhere else but found a more permanent refuge in a dead lion. A hundred years ago, a more agricultural audience would have understood honey bee processes more fully and its massive theological implications to Israel and Christ. I am afraid more modern audiences have lost much of this agricultural piece, and therefore also would not fully understood the theological impacts in the passage. Beekeepers to the rescue! Here are some references to the sweetness of honey as compared to God's redemptive love, Psalm 81:16, Psalm 19:10, 119:103, Song of Solomon 4:11, 5:1.

Finally, good stewardship is probably an easy and obvious principle for honey bees to illustrate. They are such a key piece to our ecological and agriculture systems. Proper stewardship for honey bees and the environment is top of mind for any beekeeper, but it may be easier said than done. It is a task we are challenged to take on Genesis 1:26-30, and perhaps sometimes less is more, Proverbs 25:16, 25:27.

I hope this article inspires you to take a look at the spiritual side of honey bees and how they may demonstrate the bringing peace, love, and joy into your life and the lives of others. I am currently pitching another book idea to a publisher of Bible study books to develop this course into a topical study guide to be shared with others.

The honey bee is a great gift

Bee Biosecurity

It seems there has never been a time when there has been a greater public and mass media focus on disease and sanitation. Any farm kids out there ever play "king of the mountain" on a manure pile with your cousins? I did. We didn't worry about "germs" or "bugs". When pursuing a career in studying biology, diseases, and public health, however, one comes to realize that potential infectious agents are all around us...normally. Most microorganisms are beneficial to us, but some can be pathogenic. Honey bees certainly are not immune to pathogenic and environmental outside threats. Applying biosecurity principles is one tool in the beekeeper's box that should be evaluated and employed whenever possible to prevent, reduce, and/or control biosecurity threats.

Swarms can transmit diseases

What is biosecurity for beekeepers?

Biosecurity is a set of protective protocols utilized to prevent disease, chemical, physical, and other health threats in animals and/or humans. These protocols are designed to prevent these threats from entering a population and/or the spread of disease from one population to another. Agricultural and veterinary biosecurity principles have been developed and are routine for just about every agricultural animal in the US. In animal health, biosecurity protocols are designed to avoid significant economic losses should a threat arise.

Biosecurity is considered an integral part of beekeeping in many places in the world. Australia, New Zealand, much of Europe, and Canada all have detailed recommendations and resources for employing biosecurity in their apiaries. In the US, we have room to work and improve on this. Veterinarians are extensively trained in the principles and tactics of biosecurity, which can be applied in a variety of situations.

How beekeepers may utilize biosecurity:

First, I would like to make the disclaimer that this information is **not** intended for migratory beekeepers. While certainly some of these principles could be and are applied, biosecurity is quite different for millions of hives of flying animals moving all over the country verses a stationary yard.

Also, keep in mind that **no single biosecurity plan will be practical for every beekeeping situation**. However, applying as many best practices as possible to your apiaries can reduce risks and improve your bees' health and productivity. Some plan is better than no plan.

Ten biosecurity points to evaluate in your bee yard/s:

1. **Assessment and Awareness:** Do regular hive inspections. I have talked with beekeepers who are afraid to open their hives and/or do not know what they are looking for. If that is you-stop and take some time to get comfortable. Be brave. "But what if I kill the queen?", you may ask. Well, if you do, learn how to fix it (having more than one hive helps). Learn bee biology like the back of your hand. Take beekeeping classes, find a mentor beekeeper and/or veterinarian who can help you.

    I often tell my pre-med students that one must learn what normal is before knowing what abnormal is...beekeepers must do the same. The only way to do this is to open the hive and look. During active seasons like Spring, this may mean going into your hive/s weekly. If you do not, you will miss a lot, without even knowing it, especially swarms.

    After you are comfortable with normal, be aware of possible threats (from bears to brood diseases) that could come into your yard, what they look like, and ways you could prevent them. In biosecurity speak, this is called "Hazard Identification" and "Risk Evaluation and Management". Early identification of an issue will facilitate proper intervention and/or the prevention of disease spread.

2. **Bees**: For most livestock, the introduction of new animal/s into the herd or flock is undertaken with great biosecurity precautions, as this is a major way threats can enter the operation. Purchase bees or queens from a known and trusted source. Try rearing your own bees and queens. If your bees are doing well, utilize your own stock. Once you get comfortable doing more hive inspections, you may be amazed how many split opportunities arise.

   If you manage multiple yards, keeping bees at different locations separated by 3-5 miles is a good practice, whenever possible. Not "keeping all your bees in one basket" could be critical in limiting losses should a disease outbreak or other threat arise.

3. **Equipment:** Consider the tools and equipment utilized in your bee yard. Try to develop an "all in, no out" policy for every yard. Simply having a dedicated hive tool per yard can reduce the spread of disease. Do not obtain used hive equipment, frames, tools, or wax from outside sources. All of these can harbor diseases and pests. Wax frames should be rotated out of use every three years to avoid disease transmission and pesticide exposure. Be aware that vehicles can bring pathogens and pests into a yard. Biosecurity recommendations include parking vehicles away from hives whenever possible and cleaning vehicles regularly.

4. **Sanitation/disposal:** Regular cleaning is essential for all hive equipment and tools. Ideally any wax, honey, brood, dead bees, or propolis should not be thrown on the ground but collected or disposed of away from bee access. Honey spills should be cleaned to avoid robbing behavior. Old, unusable equipment should not be lying around, but be disposed of and ideally, burned.

5. **Records:** Keeping records of your hive inspections, incoming bees/products/equipment, mite treatment, etc. is vital to tracking the health of your colonies. Providing identification for each hive (name or number) should be done. Some good record references are included below.

6. **Personnel:** People, their clothing (veils, gloves), and shoes can bring pathogens into your yard. Have visitor policies on sanitation. After all, it is your place. Ideally, in healthy yards, hand washing, or changing of gloves should be done at least between yards. Veils/jackets should be laundered regularly. If possible, set up a boot scrub area before and after entering a yard.

7. **Health practices:** Develop a regular health plan for your bees. This includes regular hive maintenance and inspections, disease monitoring (ex. routine mite counts), using medication applications properly, and an appropriate nutritional

plan. This plan should correspond to the seasonal needs of the bees. Do not feed outside honey to your bees, as honey can harbor diseases, including American Foulbrood spores. Be sure to keep vectors of disease, like rodents, in check with mouse guards.

8. **Plan in sickness and in health:** Good biosecurity ideally is meant to prevent diseases from coming into a bee yard. However, biosecurity can also prevent diseases from coming out. If an infectious disease is identified in your yard, biosecurity measures will need to be increased. Proper disposable gloves use between hives may be employed to prevent disease spread. Sick hives should be inspected last. Sick hives may need to be quarantined or euthanized. Use a dedicated hive tool or thoroughly disinfect your hive tool after use in a sick hive. Please note that an alcohol flush **will not** sterilize a hive tool. Carefully disinfecting a hive tool in a hot smoker is a good field method to reduce pathogen spread.

9. **Know how to get help:** If you suspect an infectious disease in your apiary, call for help. Know the bee laws in your state and contact your local bee inspector for assistance. Establish a relationship with a veterinarian.

10. **Continue learning:** Below are a list of resources and links that discuss bee biosecurity recommendations in greater detail. You do not have to become a biosecurity expert overnight, but practically employing these principles will reduce health risks to your bees and other bees in your community. Fall/Winter clean-up is an excellent time for assessment of your goals for the next season.

Frame marked with year to faciliate aging of frames

Another Arachnid

'Tis the season

Spring is an exciting time of renewal for nature and beekeepers. Everything is coming to life and the business of the bee yard awakens from long months of winter slumber. **Everything...** is waking up, including things we would prefer never to emerge. In the "think Spring" spirit, I would like to take some time to discuss an arachnid of concern for beekeepers... no, not Varroa mites...but ticks. While ticks do not directly affect honey bees, they certainly can affect the safety and health of beekeepers and a wide variety of our other animals. Since beekeepers spend so much time outdoors tending our bees, we are at a higher risk of encountering ticks and tick -borne diseases.

Before researching honey bees, my research involved ticks and tick-borne diseases. Since Pennsylvania has had the largest number of reported cases of Lyme Disease in the US in the last decade or so, it was a worthy public health pursuit. My research students and I worked with several PA state agencies to collect and speciate about 3000 ticks from around the Commonwealth of PA. We also tested these ticks for 5 tick-borne disease agents, *Borrelia burgdorferi* (Lyme Disease), *Borrelia miyamotoi*, *Anaplasma phagocytophilum*, *Babesia microti*, and *Bartonella*. In summary, I can tell you that we found evidence of these diseases throughout the State. We published several papers, increased tick-borne illness awareness in PA, and found the **first evidence of the Powassan virus** in the tick population in Pennsylvania.

I am happily "retired" from our tick research work and have found honey bees to be a much more attractive research subject. However, I still sit on a tick task force of such, the Tick Surveillance Community of Practice for the State. These public health meetings help to keep me "embedded" in the current status of tick-borne diseases and I would like to pass the latest information on to you.

Tick species

First, here is a quick review of various tick species that can transmit disease.

Ixodes scapularis commonly known as the Black-legged tick (locally as a deer tick, many tick species are locally known as "deer ticks," fyi), is well- established in the Eastern United States, west of the Rockies. *Ixodes pacificus* dominates the west coast of the US. And do not feel left out if you live in the Rocky mountains, Dermacentor *andersoni* or the Rocky Mountain wood tick would like to keep your company. Brown dog ticks, *Rhipicephalus sanguineus,* are distributed everywhere. There are several other species of ticks located around the country and all of them are capable of vectoring various diseases.

In clinical practice, I used to get a lot of people bringing me ticks, asking me to identify the tick, and asking if the tick could be carrying diseases (and if they could be exposed). While the identification of ticks is a fun academic practice and can have *some* historical clinical significance to disease diagnosis, the short answer here is – "Yes!" If you have been bitten by a tick, any tick, whether deer tick, dog tick or damned tick, get checked, asap.

Tick appearance and Lifecycle

Ticks have four life stages, egg, larvae, nymph, and adult. All stages, except the egg, can carry and transmit pathogens. Eggs, larvae, and nymphs are very small and can be difficult to see with an unaided eye. Nymphs are less than 2mm in size and are therefore the most common stage to parasitize humans without you even knowing it. Even adults can be tricky to find.

Ticks can have varying lifecycles; some will stay on one host their entire life, but others will use different hosts throughout their lifecycle. These three host ticks have the greatest potential to transmit diseases, since they move from host to host during their life stages. Ticks **do not** perish over winter but have lifecycles that last 2-3 years. Eggs are deposited in the leaf litter by female ticks typically in the Fall. Larvae then emerge in the Spring and typically feed on smaller animals, like rodents. This stage is typically where the ticks pick up most of their vectored diseases. Nymphs typically feed on larger animals, including humans and commonly emerge in the second Spring of the lifecycle. Fed nymphs will morph into adults by Fall, a time when adult ticks become the most active. Female adults take a blood meal from a host, mate with a male, and deposit a few thousand eggs into the leaf litter to start the lifecycle again. Both ticks die soon after mating, but if thousands of offspring are produced, one can see how an area can be infiltrated with ticks very quickly.

So, especially look for nymphs in the Spring and Summer and adults in the Fall, but host-seeking ticks can be active in temperatures above freezing all year long.

Ixodes scapularis engorged female

Tick bites hypersensitivity

In addition to vectoring many diseases, the tick bite themselves can cause a local hypersensitivity in the skin and subcutaneous tissue. Tick paralysis is a rare but dangerous neurological complication that can affect humans and most mammals. I have taught my students to add this unlikely event to their differential list. If an otherwise healthy patient presents with sudden onset of unexplained paralysis... check them for ticks.

Tick-borne diseases- Sorry, it is not just Lyme Disease

Many people and medical professionals are aware of the prevalence of Lyme Disease. According to insurance claims and the CDC, the number of people *diagnosed and treated* for Lyme Disease is 476,000 per year. They are not sure if this represents the true number since many people exposed to Lyme are asymptomatic or do not seek medical care. Also, many tick-borne diseases look clinically similar to Lyme Disease and may be misdiagnosed as such.

Your risk may vary depending on your geographical location. For migratory beekeepers, you may achieve exposure to the entire smorgasbord. Many of these diseases are bacterial in nature and others are caused by viruses. Here is a list of common tick-borne agents or diseases in the US.

Borrelia burgdorferi (Lyme Disease)

Borrelia miyamotoi

Anaplasmosis

Babesiosis

Bartonellosis

Rocky mountain spotted fever

Colorado Tick Fever Virus

Ehrlichiosis

Rickettsia parkeri

Tularemia

Similar clinical signs, diagnosis, and treatment

The typical clinical signs tick-borne diseases can cause can be remarkably similar. This can make exact diagnosis difficult. The good news is that many people exposed to tick-borne disease may never become symptomatic and clear the infection on their own. However, "flu-like" symptoms like fever, fatigue, malaise, gastrointestinal signs, headache are also quite common. Skin rashes may occur, **but they are not diagnostic**! Unfortunately, some people, including medical professionals, still believe that if a "bull's eye rash" is not found on a patient, they can rule out tick borne disease- **this is completely false!** Joint pain and neurological signs can also be found. Changes in blood work can include varied blood cell values and increases in liver enzymes.

Exact diagnosis depends on a keen medical professional picking up on your diagnosis possibilities. Incubation for many tick-borne diseases can be weeks after a tick bite, so it may be several weeks until signs show up. Also, specific and antibody blood testing will likely be negative in the initial stages of infection, so testing may need to be done weeks to months after the initial tick bite. However, treatment should be started immediately.

More, sort of, good news...many tick-borne diseases can be successfully treated with antibiotics. Doxycycline is one of the most helpful. Sound familiar? It should, as this is one of the approved and most popular antibiotics used by beekeepers on their honey bees. Tick-borne diseases are a perfect illustration of why we should use antibiotics judicially to reduce antibiotic resistance and keep doxy working for all of us. Tick-borne viral diseases are primarily treated using supportive care.

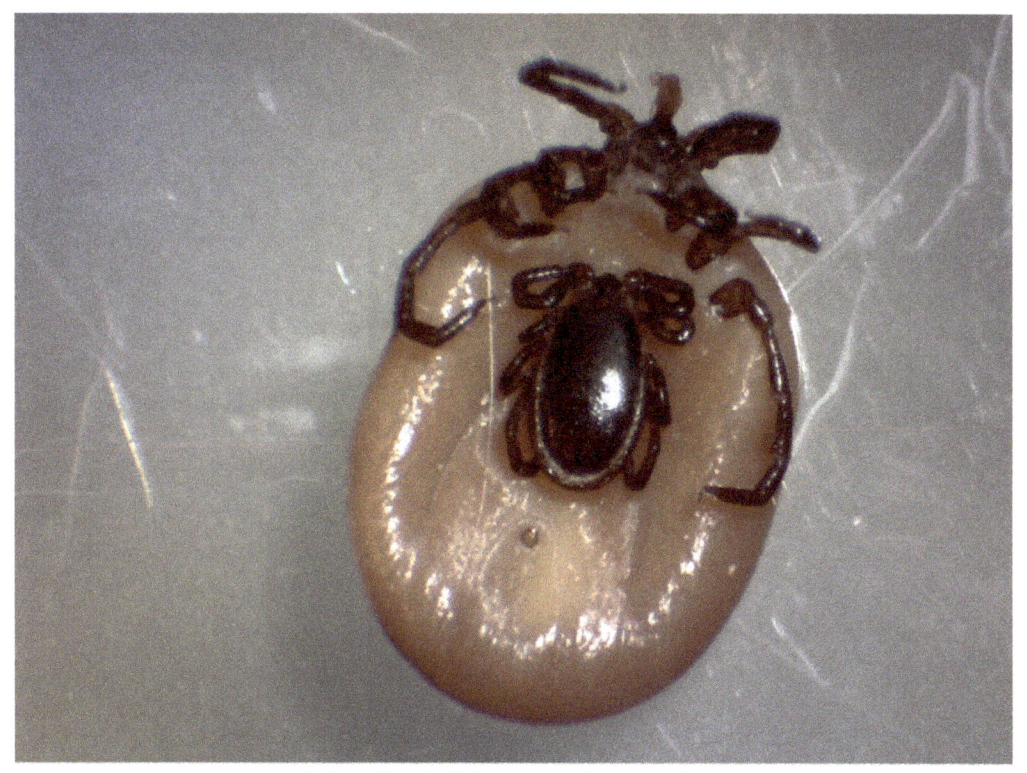

Male and Female Ticks can have thousands of babies

Preventative strategies for Beeks

Some good news... we can take several steps to reduce our exposure to ticks.

1. Preventative Sprays. Maybe you do not like chemicals or are allergic, but preventatives like permethrin applied to clothing, and EPA-registered insect repellents such as DEET applied on exposed skin are proven to be the most

effective in reducing tick exposure. Please read instructions and reapply as needed according to product label instructions.

2. Avoid tall grass and wood edges where ticks especially quest for hosts. Keeping the grass down around your bee hives is a good preventative health practice for your bees, too.

3. Ticks are less attracted to light clothing, so white clothing is good! Beeks are already ahead of the curve here. And do not forget to tuck into your boots.

4. Pets. If you have a canine friend that likes to ride along to bee checks remember, pets can pick up ticks, bring them in the house to us, and our pets can suffer from many of the same tick-borne diseases we get. Be sure to use a veterinary approved tick preventative on your pets and check over any pets exposed to tick habitats each time they return indoors.

5. Strip. This is the sexy part of the show. After returning home from the bee yard, remove all clothing, take a shower, and place clothing into the dryer on high heat to kill any lingering ticks. Examine gear such as backpacks for ticks.

6. Tick checks. Conduct a full-body tick check, including hidden areas such as the scalp, ears, armpits, belly button, and between the legs. It helps to have a partner.

7. Proper tick removal. If you happen to find a tick, do not just rip it off. Get a hemostat, tick removal key/tool, or thin tweezers to grasp the tick as close to the skin as possible and pull straight up to remove. It is possible to leave some mouth parts but just leave them alone if you do. The body will take care of it. Clean the area with alcohol, apply an antibiotic ointment, wash your hands, and consider making an appointment with your doctor. Do not use fire, teeth, petroleum products or any other kind of voodoo to try to remove a tick. Trust me, I have heard them all.

8. Find a good doc. Talk to your PCP about tick-borne diseases. Knowledgeable physicians should be happy to discuss any concerns in your area.

Wow, I know that is a lot of information. Truth is there are many disease threats that beekeepers and our honey bees face every day. But you gotta get out there and live your life! Hopefully, this chapter gives you some awareness and tools to mitigate tick-borne threat within reason and perhaps it can provide us with greater empathy to what our bees are going through with Varroa and the diseases it vectors.

The Super

I hope you have enjoyed this collection of apicultural stories from the perspective of a veterinarian, biologist, and teacher. I pray you can take away inspiration and knowledge about this lovely insect, which we have almost... tamed. The bees certainly have taught me more than I could have ever imagined! In this "super" section, I have provided resources used in the writing of this book, which I hope will be helpful to you and some "thank yous" for those who have helped me along the way.

Gifts from the hive

Acknowledgements

I would like to thank Julianna Jacobs, my research student, Megan Plunkard, my neighbor and backyard beekeeper friend, Drs. Dana Dess, Enzo Campagnolo, Jay Evans, and Monique L'Hostis my colleagues, all who agreed to read this book's drafts and provide feedback. Jerry Hayes and everyone at *Bee Culture* Magazine and Grove City College for their support in making my exploits into apiculture possible.

Normal fanning and foraging on a warm summer day

Notes/ Resources

All photos in this book are credited to and are property of the author unless otherwise noted.

Frame Two: Honey Bee Biology

Sentinels

1. Smith, K.E. Weis, D. et.al. 2020. Honey Maps the Pb Fallout from the 2019 Fire at the Notre-Dame Cathedral, Paris: A Geochemical Perspective. *Enviro. Sci. Technol.* Asc.estlett.0c00485.

2. Costa, Annamaria, Veca, Mauro, et.al. Heavy metals on honeybees Indicate their concentration in the atmosphere: a proof in concept. *Italian Journal of Animal Science*, Vol. 18, 2019- issue 1. Published on-line 24 Nov 2018.

3. Gutierrez, Miriam, Molero, Rafael, et. al. 2020 Assessing heavy metal pollution by biomonitoring nectar in Cordoba (Spain). *Environmental Science and Pollution Research* 27, 10436-10448.

4. Floyd, Mark, "Scientists use honey from beekeepers to trace heavy metal contamination." Oregon State University newsroom. March 19, 2019.

5. Zhang, Di N., Hladun, K.R. et.al. 2020, Joint effects of cadmium and copper on *Apis mellifera* foragers and larvae. *Comparative Biochemistry and Physiology Part C: Toxicology & Pharmacology 237*, 108839.

6. Szlard Bartha, Taut, Ioan, et.al. 2020. Heavy Metal Content in Polyfloral Honey and Potential Health Risk. A Case Study of Copsa Mica, Romania. *International Journal of Environmental Research and Public Health, 17, 1507; doi:10.3390/ijerph17051507.*

When Considering the Data

1. https://health.clevelandclinic.org/body-temperature-what-is-and-isnt-normal/ Accessed 12/09/2021.

2. https://wexnermedical.osu.edu/blog/new-normal-body-temperature Accessed 12/09/2021.

3. https://www.health.harvard.edu/blog/time-to-redefine-normal-body-temperature-2020031319173 Accessed 12/09/2021.

4. https://www.psychologicalscience.org/news/repeating-misinformation-doesnt-make-it-true-but-does-make-it-more-likely-to-be-believed.html Accessed 12/09/2021.

5. https://www.beepods.com/honey-bees-survive-winter-regulating-temperature-cluster/ Accessed 12/09/2021.

6. https://www.ncbi.nlm.nih.gov/books/NBK263242/#S23 Accessed 12/09/2021.

Data pyramid source: https://latrobe.libguides.com/ebp/study-design Accessed 12/10/2021.

Farone, Tracy S., Registered Medicinal Products for Use in Honey Bees in the United States and Canada, Veterinary Clinics of North America: Food Animal Practice, Volume 37, Issue 3, 2021, Pages 451-465.

The Hexagon, Under the Microscope

More examples of histological tissue discussed can be found here:

Full urinary bladder lumen. https://histologyguide.org/EM-view/EM-238-bladder-stretched/16-photo-1.html. Accessed 03/29/2022.

Liver lobule under a light microscope. http://www.vivo.colostate.edu/hbooks/pathphys/digestion/liver/histo_lobule.html. Accessed 03/29/2022.

SEM of small intestinal lumen. https://www.cram.com/flashcards/histology-of-gi-tract-3823044. Accessed 03/29/2022.

Lens fiber SEM. https://basicmedicalkey.com/crystalline-lens/. Accessed 03/29/2022.

FS in Nosema detection. https://pubmed.ncbi.nlm.nih.gov/30719512/ accessed March 28th, 2022.

Frame Three: Honey Bee Anatomy and Physiology

*Dr. Farone has taught Anatomy and Physiology courses for 25 years. Much of the information shared here comes from material found in her courses and abundantly throughout the literature.

Time for Some Muscle

Anything written by Snodgrass.

Ramsey SD, Ochoa R, Bauchan G, Gulbronson C, Mowery JD, Cohen A, Lim D, Joklik J, Cicero JM, Ellis JD, Hawthorne D, vanEngelsdorp D. *Varroa destructor* feeds primarily on honey bee fat body tissue and not hemolymph. Proc Natl Acad Sci U S A. 2019 Jan 29;116(5):1792-1801. doi: 10.1073/pnas.1818371116. Epub 2019 Jan 15. PMID: 30647116; PMCID: PMC6358713.

Great anatomy pictures and muscle views. https://bee-health.extension.org/anatomy-of-the-honey-bee/

Vidal-Naquet, Nicolas. Honeybee Veterinary Medicine: *Apis mellifera* L., 5m Publishing, 2015.

All Things Change

https://bee-health.extension.org/head-segment-of-the-honey-bee/ accessed 10/03/2023.

Menzel, R. The honeybee as a model for understanding the basis of cognition. *Nat Rev Neurosci* **13**, 758–768 (2012). https://doi.org/10.1038/nrn3357 accessed 10/03/2023.

https://askabiologist.asu.edu/honey-bee-anatomy accessed 10/03/2023 -a fun bee anatomy quiz.

https://bees.msu.edu/honey-bee-anatomy/ accessed 10/03/2023.

New Beginnings

1. Collison, Clarence. "A Closer Look: Queen Post-Mating Changes." *Bee Culture Magazine*. April 2021. Pp. 32-34.
2. Collison, Clarence. "A Closer Look: Queen's Reproductive System." *Bee Culture Magazine*. February 2022. pp. 32-34.

Collison, Clarence. "A Closer Look: Honey Bee Eggs & Oogenesis." *Bee Culture Magazine*. December 2020. pp. 33-36.

Collison, Clarence. "Drone Biology and Behavior." *Bee Culture Magazine*. March 2021. pp. 56-59.

Vidal-Naquet, Nicolas. Honeybee Veterinary Medicine: *Apis mellifera* L., 5m Publishing, 2015.

Beeinformed https://beeinformed.org/wp-content/uploads/2023/06/BIP-2022-23-Loss-Abstract.pdf

Accessed November 1, 2023.

More on transgenerational priming: https://www.dalan.com/media-publications

In with the New and Out with the Old

Collison, Clarence. "A Closer Look: Digestive and Excretory Systems." *Bee Culture*. August 2020. pp. 35-7.

Bee Beats and Breaths

https://beekeephub.com/the-circulatory-system-of-bees/ Accessed 01/24/2023.

https://americanbeejournal.com/the-internal-anatomy-of-the-honey-bee/ Accessed 01/24/2023.

https://americanbeejournal.com/respiration-and-circulation-in-honey-bees/ Accessed 01/24/2023.

Frame Four: **Honey Bees Behavior**

Human-Bee Bond?

Honeybee memory: a honeybee knows what to do and when

Shaowu Zhang, Sebastian Schwarz, Mario Pahl, Hong Zhu, Juergen Tautz

Journal of Experimental Biology 2006 209: 4420-4428; doi: 10.1242/jeb.02522

Biergans SD, Claudianos C, Reinhard J and Galizia CG (2016) DNA Methylation Adjusts the Specificity of Memories Depending on the Learning Context and Promotes Relearning in Honeybees. Front. Mol. Neurosci. 9:82. doi: 10.3389/fnmol.2016.00082 http://journal.frontiersin.org/article/10.3389/fnmol.2016.00082/full

M.E. Villar et al., "Redefining single-trial memories in the honeybee," *Cell Rep,* 30:2603–13.e3, 2020.

The Human-Animal Bond Research Institute: https://habri.org/

AVMA's definition of the human animal bond. https://www.avma.org/one-health/human-animal-bond

I found one master's thesis on the subject of the human-animal bond and bees: Accessed 09/07/2020 https://aquila.usm.edu/cgi/viewcontent.cgi?article=1213&context=masters_theses

The Conundrum of Feral Honey Bees

1. "Feral." *Merriam-Webster.com Dictionary*, Merriam-Webster, https://www.merriam-webster.com/dictionary/feral. Accessed 22 Nov. 2022.

2. "Human Relationships with Honey Bee date back 9,000 years." https://www.pbs.org/newshour/science/humans-relationship-honeybees-goes-back-neolithic-era Accessed 22 November 2022.

3. Roffet-Salque, M., Regert, M., Evershed, R. *et al.* Widespread exploitation of the honeybee by early Neolithic farmers. *Nature* **527**, 226–230 (2015). https://doi.org/10.1038/nature15757. https://rdcu.be/c0dtC Accessed 22 Nov. 2022.

4. "Sylvatic." *Merriam-Webster.com Dictionary*, Merriam-Webster, https://www.merriam-webster.com/dictionary/sylvatic. Accessed 22 Nov. 2022.

5. Rax, David. "A Brief History of House Cats." Smithsonian Magazine. June 30, 2007. https://www.smithsonianmag.com/history/a-brief-history-of-house-

cats-158390681/ Accessed 22 Nov. 22.

6. USDA 2021 Honey report. Released March 18, 2021. https://downloads.usda.library.cornell.edu/usda-esmis/files/hd76s004z/7m01cp956/df65wc389/hony0322.pdf. Accessed 23 Nov 22.

7. "Number of Beehives in Leading Countries Worldwide in 2020. https://www.statista.com/statistics/755243/number-of-beehives-in-leading-countries-worldwide/. Accessed 22 Nov. 22.

8. Life Expectancy in the USA 1900-98. https://u.demog.berkeley.edu/~andrew/1918/figure2.html. Accessed 22 Nov.22. Also interesting but not referenced US life expectancy stats: https://www.simplyinsurance.com/average-us-life-expectancy-statistics/

9. Africa.com. United Nations Report. "Healthy Life Expectancy grows by 10 years". https://www.africa.com/healthy-life-expectancy-in-africa-grows-by-nearly-10-years/ Accessed 22 Nov.22.

10. Honan, Kim," Feral Honey Bees to Be Poisoned in NSW Varroa Mite Hotspot." Oct.5, 2022. https://fans2pets.com/feral-honey-bees-to-be-poisoned-in-nsw-varroa-mite-hotspots/ Accessed 22 Nov.22.

Frame Five: Honey Bee Management

Veterinary Diagnostic Approach

The Diagnosis

Honey bee laboratory resources for confirmatory testing:

https://bee-health.extension.org/usda-ars-bee-labs/

https://www.ars.usda.gov/northeast-area/beltsville-md-barc/beltsville-agricultural-research-center/bee-research-laboratory/

https://www.ohiostatebeekeepers.org/2017/announcements/news/nagc-launches-beecare-testing-for-honeybee-diseases/

https://www.gov.mb.ca/agriculture/animal-health-and-welfare/vds/pubs/vds-lab-manual-honey-bee.pdf

https://beeinformed.org/

Treatment

1. Zoonotic disease prevalence: https://www.ncbi.nlm.nih.gov/pmc/articles/PMC5711306/#:~:text=Emerging%20and%20endemic%20zoonotic%20diseases,origin%20(1%2C2).

2. Human cases of *P. larvae* bacteremia: https://www.ncbi.nlm.nih.gov/pmc/articles/PMC3322038/

3. CDC information on antibiotic resistance: https://www.cdc.gov/drugresistance/solutions-initiative/stories/ar-global-threat.html

4. WHO information on antibiotic resistance:

https://www.who.int/news/item/29-04-2019-new-report-calls-for-urgent-action-to-avert-antimicrobial-resistance-crisis#:~:text=By%202030%2C%20antimicrobial%20resistance%20could,die%20from%20multidrug%2Dresistant%20tuberculosis.

Veterinary approved bee antibiotics: http://www.farad.org/vetgram/honeybees.asp

Great resource for best practices and management of varroa and other diseases:

https://honeybeehealthcoalition.org/

If We Only Followed the Directions!

1. Panesar, Kim. "Patient Compliance and Health Behavior Models", *U.S. Pharxmacist*. April 23, 2012. https://www.uspharmacist.com/article/patient-compliance-and-health-behavior-models accessed 02/01/2021.

2. Little, Geoff. "Concordance and Compliance." *Veterinary Practice*. Feb. 2013. https://veterinary-practice.com/article/concordance-and-compliance accessed 2/01/2021.

Frame Six: Honey Bee Medicine

Diagnosis Is a Tricky Thing

Randy Oliver developed a nice loop diagram showing how different underlying causes of colony loss can intermingle and have compounding effects.
http://scientificbeekeeping.com/sick-bees-part-2-a-model-of-colony-collapse/

Euthanasia

1. https://www.avma.org/sites/default/files/2020-01/2020-Euthanasia-Final-1-17-20.pdf Accessed 09/09/2021.

2. Kane, Terry R. and Faux, Cynthia M, ed. "Honey Bee Medicine: For The Veterinary Practitioner". Wiley Blackwell. 2021. pp.356.

3. Honey Bee Veterinary Consortium – How to Euthanize a Hive. https://www.hbvc.org/content.aspx?page_id=274&club_id=213546 Accessed 09/09/2021.

Immunity, Vaccines & Honey Bees

Freitak, Dalial, Schmidtberg, Henrike, Franziska, Dickel, et.al. The maternal transfer of bacteria can mediate trans-generational immune priming in insects. Virulence. 5:4.547-554. May 15, 2014.

Freitak, Dalial. "Vaccination of Honey Bees Against American Foulbrood." *Bee Culture*. Feb 2021. pp. 76-7.

Goblirsch, Michaeal and Freitak, Dalial. Proceedings of the 2020 American Bee Research Conference

2.20. Testing a Queen Vaccine against Chalkbrood Infection (Poster).

Harwood, Gyan, Amdam, Gro V., Freitak, Daliel. The role of Vitellogenin in the transfer of immune elicitors from gut to hypopharyngeal glands in honey bees (*Apis mellifera*). *Journal of Insect Physiology*. 112(2019) 90-100. doi:10.1016/j.jinsphys.2018.12.006.

Harwood, Gyan, Heli Salmela, Freitak, Dalial and Amdam, Gro V. Social immunity in honey bees: royal jelly as a vehicle in transferring bacterial pathogen fragments between nestmates. Journal of Experimental Biology. (2021) 224.jeb231076. doi: 1242/jeb.231076.

Heli Salmela, Amdam, Gro V., Freitak, Dalial (2015). Transfer of Immunity from Mother to Offspring is Mediated via Egg-Yolk Protein Vitellogenin. *PLoS Pathog* 11(7):e1005015. doi:10.1371/journal.ppat.1005015.

Hernández López, Javier, Schuehly, Wolfgang, Crailsheim, Karl, and Riessberger-Gallé, Ulrike. Honeybee transgeneraltions immune priming using injection of Paenibacillus bacteria.

https://www.ncbi.nlm.nih.gov/pmc/articles/PMC4024302/

Frame Seven: Honey Bees Diseases

Tropilaelosis

De Guzman, Lilia I., Williams Geoffrey R., et.al., "Ecology, Life History, and Management of Tropilaelaps Mites", *Journal of Economic Entomology*, Volume 110, Issue 2, April 2017, Pages 319–332, https://doi.org/10.1093/jee/tow304 Published: 08 March 2017.

Vidal-Naquet, Nicolas. Honeybee Veterinary Medicine: *Apis mellifera* L., 5m Publishing, 2015, pp.138-142.

OIE policies on Tropilaelaps mites:

https://www.oie.int/fileadmin/Home/eng/Health_standards/tahc/current/chapitre_tropilaelaps_spp.pdf.

Great picture of Tropilaelaps:

https://beeaware.org.au/archive-pest/tropilaelaps-2/#ad-image-0

USDA bee mite ID, Tropilaelaps:

http://idtools.org/id/mites/beemites/factsheet.php?name=15241

Tropilaelaps info sheet:

https://www.aphis.usda.gov/plant_health/plant_pest_info/honey_bees/downloads/Tropilaelaps-InfoSheeta.pdf

USDA national honey bee survey information including surveillance for Tropilaelaps:

https://www.aphis.usda.gov/plant_health/plant_pest_info/honey_bees/downloads/SurveyProjectPlan.pdf

USDA reportable bee diseases:

https://www.aphis.usda.gov/aphis/ourfocus/animalhealth/monitoring-and-surveillance/sa_nahss/status-reportable-disease-us/!ut/p/z1/lZJNU4MwEIZ_Sw8cIRvaodQbIFNQqGMtirl0gqbADCVMEmT015vWkx9tMZd8zPPuZt9dRFCOSEvf6pKqmre00fdn4mxXOFiCO8PJMrzG4EWL-DaaOwD3Nno6AsndLMD-A-g99MEL1_M0DGMb8BSR_-kfI0frN9lm5WI